MICROCLIMATES

taylor brady

KRUPSKAYA • 2001

ACKNOWLEDGEMENTS

Sections of this book have appeared or are forthcoming in *Berkeley Poetry Review, COMBO, Kenning, Mirage #4/Period(ical), The Queen Street Quarterly, Ribot,* online at the *Narrativity* website (www.sfsu.edu/~newlit/narrativity), and in the anthology of the 2001 Euro-San Francisco Poetry Festival, published by City Lights Books. My thanks to Amar Ravva, Mike Magee, Patrick Durgin, Dodie Bellamy, Kevin Killian, Stephen Cain, Paul Vangelisti, Camille Roy, Bob Glück, Mary Burger, Steve Dickison, and Hugh Steinberg for their editorial support of this work, and their efforts to make a place for poetry and exploratory narrative among the discourses of the contemporary. Special thanks to Renee Gladman, whose Leroy Books published the first section of the prose herein as a chapbook, *33549*. Her care and generosity as editor, booksmith, reader, writer and friend are deeply embedded in *Microclimates*, as are the work and conversation of Tisa Bryant, Brent Cunningham, Rob Halpern, Jen Hofer, and Laura Moriarty, among others too many to name. The musics of Anthony Braxton proposed with joy and clarity some of the possibilities for the overlap of multiple formal and performative logics that the book hopes to explore: uncited in the text, his work nonetheless sounded in my room and in my head throughout the writing. Finally, Tanya Hollis commented early, often, and acutely on the various states of the manuscript in progress. Her ability to read in a mode at once demanding and encouraging is one of many kindnesses for which I continue to thank her.

Cover Art, "Index Print," by Tanya Hollis
Cover Design by Frank Mueller

Distributed by Small Press Distribution, Berkeley
800-869-7553
orders@spdbooks.org

ISBN 1-928650-09-0

KRUPSKAYA
PO Box 420249
San Francisco, CA
94142-0249
www.krupskayabooks.com

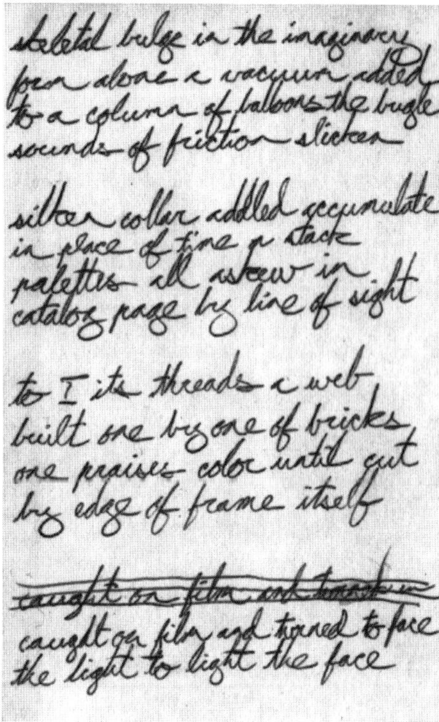

skeletal bulge in the imaginary
fern alone a vacuum added
to a column of balloons the bugle
sounds of friction slicken

silver collar addled accumulate
in place of time a stack
palettes all askew in
catalog page by line of sight

to I its threads a web
built one by one of bricks
one praises color until cut
by edge of frame itself

~~caught on film and turned~~
caught on film and turned to face
the light to light the face

[1]

[1] [Eds.] In the manuscript this space was occupied by a Polaroid photograph set between corners. Its position in the front matter warrants some speculation that it presented a portrait of the author. Unfortunately, severe water damage – perhaps inflicted by the events narrated in pp. 63-73 – has rendered the photograph unrecognizable. As we cut the smudged image away from the page, however, we took note of a short handwritten text on the reverse, apparently protected from the deluge by the glossy image side. Thus, the portrait gave its life to save what we cannot help but read as its own derivative commentary, which we present here in lieu of the portrait itself.

33549

The characters of the time seem strangers to the characters of the lightning.
— Louis Althusser

I have been known to sleep at high velocity, rounding corners into that historical narrative, written in the third person, in which the pronoun "I" names a character reading its own biography into a slow wakefulness of overture, where sovereignty turns on relations of indifference with rule in a black box left uncontemplated. The softest objects there have a kind of face, while my own face manages only to approach itself from afar in a comfortable world of hurt. And then always on this side of the wall, I have found that furniture, piece by piece, from which I differed only by a slight degree in dreams I "had" by refusing sleep and the long, slow, imaginative fuck it offered. With such advanced technologies at my disposal I controlled the weather and the dispositions of astronomy and physics, dying repeatedly of damp and vacuum. "The clothes make the man" was my motto along with everyone else, but one did well to remember that others wear those silks, linens and synthetics on the scale of epochal costume drama, while one's own body is simply too busy recreating room and rooms for itself to be bothered with more than a shapeless drape or fall in muslin or mylar. This vagueness was always the exact size of the present tense and its culpable ownership of certain craft objects in stolid, smoky glass, suspended heavily from the center of its dome. I found my body finally in that attitude, a chain stapled at one end to the hard transparence of the sky's obscuring of its own far side, threading its full length in constant recomplication through the world in the manner of a promise that at every post where *something* stood, there also would be a knot, binding up and squeezing like a muscular snake in its indissoluble folds lasting the entire space of a single night those bits of velveteen, wool, mahogany, yellowed paper of hackneyed pirate adventures, and swept wings of certain tiny birds whose delicate and durable spittle made stony edifices fading always farther into beadlike permanence, strung along that selfsame chain always taken up in relaxing its each sore clench in turn so as to come itself into the foreground in a stroboscopic photograph of nothing moving very much at all. Someone in that place was always counting, "one, two...twelve," you could see it on her face and hear it loud inside her when the grownups downstairs stopped talking, the inexorability of the sequence ignoring and nullifying me even as the enunciation of each number in turn sounded the dimensions of that chamber whose size and shape an upward geologic pressure urges me to this day to become. It marked a limit to my own talents of perception, my own ability to have a clear and distinct idea of the political architectures in the intimate air around me, that I remained undecided whether all this furious, vague movement marked a flight from successive generations in the images of women's bodies, or took place entirely within them, as in the example of Bergman's rear-projected tarantula providing both the icon and the curtain for childhood. Just then, words began to push worlds from my mouth, sciences of prediction colored and weighted like money jingling at the back of a seldom-opened drawer. Each such world was a ballooning crop of bone pushing through the pastoral sward, a skeletal bulge in the imaginary, around which thin, membranous bodies would wrap themselves as proxy faces for my own bad habit of aversion.

* * *

But time enough for all that earlier: did I mention that Pop did the weather report, and that rain fell, welcome or not, on at least two other members of our family, saturating their skins all through this dissolute late middle age of our lives together, when I was still barely tall enough to stretch into the view from the lowliest of windows? Grandma Violet threatened often to return to mud altogether, her eyes become wet lips to call for help and pleasure, held out above the bog's advancing suck as a wry emergency of puckered flesh. At the opposite end of the house from the grandfather who lacked both name and second nature, I arrived too early on the hidden periphery of this scene by jerking off against a blank, white wall, letting what few drops were spilled there develop through the force of my own melancholy pre-teen gravity into anything but the writing an unconscious and transoceanic plagiarism demanded they annul by becoming, and thus refute itself.

One day, at last, there was the event of another mouth, and then always after that was the absence of that mouth, so that your wetness on my skin always signs that future tense in which you shall have been long dead, or else long occupied in ladling up soup for the others whose voices I can hear through the floor vents, but whose faces never appear. Nonetheless, something on that day rang a bell, and the flora came to attention as the fauna pretended not to notice, except in the offhand way in which one takes cognizance of the distinction between a pocket full of change and a skillet full of eggs. Anyway, we kept the plants in the dark as much as possible. Was this the same darkness of which our sudden visitor's father had once spoken, that old water bird gone beakless on the cropped lawn, when he confessed a dazzling intermittence of light that, if not night, was just as blind and far less comfortable, glints of gold intensified by their minor interval of duration occurring as a caution to the historical narrative that would put its eggs in a single basket, or any basket at all, changing its pocket full of change for a head full of changes? In any event, his son was taking a different train these days, despite our nearly ontological commitment to the immutability of transit. That afternoon, as I emerged from my slightly stained attic room into the wet sunlight of the hour following heavy rain, I caught wind of him deflecting the drift of such things by talking at length of walls and angles – not a subject I was especially inclined to share with the rest of the assembled family. Apparently the thoughts of my Auntie Trifle ran along a similar track, for just then she saved my face by re-opening, in the most vulgar manner possible – all eyebrows and ingratitude – the question of this Mr. Bird-in-Your-Hand's rather too intimate relationship with the railyards and their stack of questionable boxes. If it was art, she reasoned, then she was all for eating it up and getting it over with once and for all, if only to stop him sneaking off into his hollow head, or at least to find out where it was exactly that he went at times like this.

"But my dears, I am right here in your company," his voice would hail us, a form alone wafting over the content-rich afternoon, and we would obligingly fill in the details, which were plentiful enough given the setting. Listening to the grass grow, I thought then to myself, as all ten-year-olds must from time to time, "His look is my foolish youth of long ago, a vacant room in which I'll squat awhile."

And so, adjusting my ears a little, I prepared myself to hear him speak his judgment from some other place altogether, as the expensive sound system spatialized the late afternoon for us, hard-panning left and right in rapid and arrhythmic alternation: "There are no gifts among you people, only a potluck to which each guest brings a dish of home-cooked debt to pass, a nightclub in which the dancers take turns wearing some flawed article of clothing to provoke the others into ecstasies of mending. I assume my presence is simply what you need in order to differ from yourselves, as I am the shock of your own names in all the papers. And you, little man, what wet flesh are you holding there under your clothes?"

It will perhaps stretch my credibility in your eyes, and thus distort your entire field of vision herein, dear reader, to protest against an easy assumption so early in the course of our travels together, but I must maintain nonetheless and absolutely that my reaction to this question, or rather, my failure to react, was motivated by nothing like a paralyzed mortification in the face of this naming my clandestine pleasures into full view, but entirely by the unassailable conviction that I knew on which side my door was closed, and was in fact present always in mind at its closing, listening somewhat fanatically to the satisfying click, feeling the slight wobble of the hollow-core slab as it settled into the cracked frame, and that knowing myself thus positioned, the thought that any might address me from so remote a location as the one Mr. Bird-in-Your-Hand occupied at our table, which was, strictly speaking, not even at the table but in the gradual understanding that he was always in a condition of gently surprising us with his arrival at a gathering already in progress, and that we were thus free to leave him seated there smoking a thin joint and drinking his slowly warming beer in order to go off in search of him – for surely, he was set to emerge into the yard from around the corner of the privacy fence at any moment – this thought, I must insist, never occurred to me. An unexpected silence in the flow of the music at precisely this moment was in fact what caused me to jump, and Grandma Violet's left-hand shadow to disappear for a moment altogether, as if perhaps a question rhyming with the question which, as we have seen, could not have been addressed to me by any stretch of the imagination, had nonetheless discovered all of us exactly where we stood, in the attitude of having been terrorized by its intonation over the course of years, exchanging threadbare pleasantries and obliging gestures as we waited, never breathing, for it to depart from the suburb where, appearances of the equipment to the contrary, our means remained quite modest. It was only that certain consumer goods were indispensable adjuncts to any departure from an overweening respectability, as when Auntie Terrible – who was Auntie Trifle's younger sister, a deliverywoman and writer of crank pamphlet literature on the proper way to put your hands inside a dead person – was arguing the merits of shaking children rather than bottles, it was the bottle, full if not entirely expensive, that was nonetheless passed around for the purpose of demonstrating her thesis, while I was left rather studiously alone.

"Let us study how best to leave the child alone," my mother in fact suggested, "in keeping with our preference for the intensive over the extensive popping off the occasional candid snapshot – here he is up a tree, here he is dancing nude in a straw hat, here he is muffled in his father's bulky headphones – and filling the long blank stretches in between with an expressionistic smear of the excess energetics spilling from the strait confines of those flashlit, redeyed moments, naming the whole assemblage a film. This membrane will separate the rhythm of our recollections from the oppressive odors of the brewery venting its yeasty gases and the orange grove burning its winter smudgepots, and allow us all a heroism without the usual dull and nagging throb. No, don't write me when I'm improvising lines," with this turning towards me in my first ever experience of direct address, timed to a different tense altogether, a kind of science fiction.

"Respect the integrity of the residue of this not-quite-primal scene as you would the bodies of the dead, which is to say," – and here she cast a baleful look at Auntie Terrible, with whom this was a constant bone of contention – "keep your hands out of my interior, since you have only slightly less claim to it than do I, who am little more than a bit of vacuum added to the neutral plane of your own blank refusal to recognize who or what I might have been aside from an exemplary pedagogy, an instructional film which infrequently, as a reward for some especially laudable bit of behavior or other, you and your comrades will be allowed to view at high speed and in reverse, this labor of reproduction, respooling the punishing lesson for the next repetition of the daily blow by blow, counting as a signal pleasure in your regulated libertine pursuits. Haven't you imagined that there are recesses, even shallow mass-produced crap like this bowl of nuts," striking her fingernail against the dull, salt-caked plastic, "into which you cannot reach, just as none of us has managed to figure out what depth that sweet and acrid stink of burning really rises from? I can see you in that soupy future tense – even long dead I can see you, telling how you felt all this so deeply – but here and now I tell you, in fact I insist, and you will memorize it: you feel nothing, but act within the anaesthetic trance of a command to learn. Get close, get real gone – real, then gone. No one's going to speak to you only with her eyes, even if that's the only way you'll listen. Here's your deal: soft blowing on soprano sax in the next room, no pitch, just breath over the reed, sounding against your closed lids like a distant music trying to sound like a close-up sound sounding far away. School tomorrow will punish you terribly for imagining that I existed. No sliding panels, only corridors, and an indifferent mass of siblings who have your body to share among them. Or bottles, I forget which, now that Terrible has authored this deadly confusion between counting money at the dinner table and having one word follow another when we talk.

"But your fear of the dark is a nightlight for both of us, as light must follow darkness, as father and stepfather must follow us upstairs, as two stories must follow the split-level which follows the flat-roofed cinderblock duplex in our narrative of just how big a tract history can occupy. And your fathers are nothing if not intractably tractable, working as they do in the absence of revolutionary theory and given over in full to spontaneist 'actions' to convert going up in smoke into a column of balloons celebrating strict release, totaling zero in a perfect act of balance. If you learn to turn the volume down on a world full of houses full of children being beaten, you can hear the boy down the block, out in the street, being clubbed by cops to whom he has been sacrificed, soft sound of blows echoing back to us through the empty storm drains, which is perhaps our own best attempt at 'functional density' triggering 'programmatic instability' in the political architectures close around us in the intimate and humid air. So have a little aleatory and shut up – this is birdsong! You, now called adolescent and scab, have junked my judgment for a boss to rebuild judgment on us all. Now you're working, read the manual, really very lovely, nothing prior to this moment has existed, yum yum, glub."

Mom's voice was occupied at least in part with the possibility – some large proportion of the partials making up the positive I.D. of its unique timbre being in fact entirely given over to it – that it had once belonged to someone else, and was now swallowed up or submerged in a collection present relations had assembled in a hypothetical catalog or how-to manual for the life I might have led, having been born in the relaxed grip of evidence in a brushed concrete, five-story postmodern airport parking structure, in the impossibly heat-hazed and blue-shifted distance of the impossible year 1970, rather than – as habitually was the case – the slippery squeeze of fiction into a mouth first tasting, next drowning in the thickened January air around the functionalist brick pile of the county hospital in 1972, which despite Mom's progressivist narrative construction work was the only time in the next twenty-five years she'd live farther from ground than a cracked, terrazzo-topped slab, the year itself being situated at an intersection where, notwithstanding memorial plaques affixed to built environments not so much looming as escaping toward their vanishing points on the three remaining corners, things neither began nor ended, but faltered just a little. Perhaps any narrator only manages to distinguish a body from the furniture by crashing through the worn pressboard seat, improvising ass first a gestural series of responses in the wake of getting quickly down to ground, breach-birthed out of memoir and portraiture into commerce and the realtor's flash photograph.

"Teach the boy to read in reverse," Grandma Violet burst in, the bugle of her voice interrupting with a call to arms these projections of narrative time onto a wall as yet unbuilt but wholly real, "and he'll know better where you're coming from, what long-demolished drywall is still punctured by the tack that holds your overexposed yearbook pic."

Everything left out returned to pile ruddy flesh onto these equivocations. Mom, following Grandma's advice, fed me whole grammars of sesame-and-honey candy – the sweetest substance foodstamps could buy – and let me practice a cheater's divination, in which the digestive action of each moment upon the discount sneakers, alphabet flashcards, and plastic dolls consumed a breath or two before produced the warm embrace of my full relation to my time. There was, to give only the most immediate example, a little girl waiting under the banana tree around the chain-linked corner, knowing I would have to kiss her sooner or later, blown backward as I was towards that future – into which she was peering to arrive through thick plastic glasses – by a hot wind from heaven, or if not heaven, at least the not-yet-defunct port just past downtown, full of coffee, phosphates, thick transparent plastics used to make cheap lenses, and the wage labor for which in some sooner future a vaguely-named and later "we" would inhabit the command to learn to be grateful in retrospect as if for the temperate bounty of the Gulf of Mexico itself – this wind, wherever from, caught in my outstretched and non-angelic fingers still rigid now ten years along with that inaugural red-faced squall against which all the variations on the family monologue appeared to me as if arrayed in an attempt to make me breathe, speak, or just shut up and *be*.

I had become, in the course of taking on this sweet and sticky ballast, the kind of kid who wrote letters to men doing time, and who heard trees whispering his name in answer as he passed through the over-irrigated landscape, still unable to speak how straight my head had got it all.

"Hey kid," one of them said to me in a tone of insinuation and minor threat, "that thumb in your mouth is just the tip of a universal iceberg, unrelated to the scale of your own remaining years," – which I ticked off in advance with a pen knife in the thick bark that screamed but never suggested that I stop. Information fell through waxy leaves onto my tongue the same as ever, but gradually less and less. I understood then, hearing the slow voice inside the wood, that this particular oak had traveled great distances to tell me it was dead, and that people live inside things like iron filings in food, a fairly serious irritant – and that this would make a tiny root pierce my heart and a bare, milled trunk erupt from my body, wires strung from its level top calling out to all the others who were likewise consumed and paralyzed while the trees went around hopping mad like Mom when she stubbed her toe, trying to extract the splinter-people embedded in their sensitive pulp, and we wondered whether these prisoners were all growing tiny phone poles of their own. Each nest of wires nestled other tangles, and the distributions of town and country grew from the tingle and flow of that brain-damaging electromagnetic field.

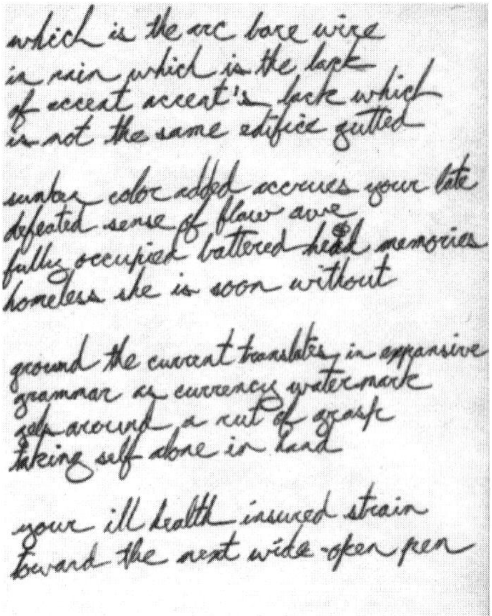
which is the arc bare wire
in rain which is the lack
of accent accent's lack which
is not the same edifice gutted

water color added accrues your late
defeated sense of flaw awe
fully occupied battered head memories
homeless she is soon without

around the current translates in expansive
grammar as currency watermark
seep around a cut of grasp
taking self alone in hand

your ill health insured strain
toward the next wide-open pen

We believed in the city throughout those days, which is to say we did not quite live there, but made out the mark it made, out against the scrub palmetto, at least on smogless days, and took it for a sign.[2]

[2] [Eds.] Unless otherwise noted, all photographic inserts in the manuscript found in sections written before the flood on pp. 63-73 are in similar condition to the author portrait. The reader will note that in the present instance, as well as in several others, someone seems to have confirmed in advance our decision to present these short poems as impoverished images by captioning them as if they were, in fact, photographic plates. The hand is not the anonymous author's, and there is some dispute among ourselves as to whether the captions were present when we first received the manuscript and cut the damaged photographs loose.

A road came halfway forward from a disavowed distance, then fell apart into a tangled nest of smaller roads that sprayed the stretch of tract housing, in which we all worked at one time or another as landscapers, with the smell of a tight-packed, large-scale life that was, at this distance, microbial and decorative, infinitely broader but infinitely flatter than the atmosphere to which it gave rise in our noses sniffing out the blood-piss-and-worn-coin tidal stink of the large. This was the undertone and undertow in each lifted spoon of lime gelatin with miniature marshmallows raised to my mouth, Home Ec in every way conceivable. Against which stood, or rather lay, Auntie Terrible – her name is my name too – whose every artifact smelled of the undercooked bread she handled on her delivery rounds, whose eyes were always opening, who felt everything and savored almost nothing, who had things like ball bearings in her brain that made her rattle like a dying person when she lay, never forgetting to stare out from her face, unsleeping on principle.

"Insect, discount merchandise, dismembered flower," she rattled on and on, building homes for mockingbirds out of these possible names for what was loose inside her head. Everything was always happening at the same time, so we saw each other dead as often as alive, leaves blown to ash off trees blown down like shingles blasted from test-range houses in high-speed film studies, and Terrible and I – who bore an edge of flashing from the same injection mold – would often eye each other knowingly before scuttling off for the scheduled dose of ginger ale, which Grandma Violet insisted would maintain the gastronomic conditions for sustained belief in the fiction of a polis, a politics, even in the face of such precision-timed uncertainty. A smile on the lips of Terrible – who had my number – would break my face from time to time. An assistant, Americana, helped her load and drive her truck, but found it more amusing to hang around and cook weird meals in my mother's kitchen. On one such occasion, I sidled up to her to hear the story of a machine for looking at your own heart as it squelched against your insides, which seemed to resonate in an oddly archaic way with Terrible's professed interest in the cavities of the dead – though when I mentioned this to Americana, she shushed me in a manner simultaneously self-deprecating, dismissive, and dirty. Nothing really shocked her, though. Even my periodically shattered smile occurred to her as part of a quotidian ambience, like sunlight, genetically modified vegetables, and outmoded coal-burning electrical generators, smoking into thick air down there on Cockroach Bay.

New children were always exciting, though, since you couldn't yet compare them to men. I think this was why, on an afternoon after that first occasion of rain, but before a rather more definitively wet episode, I caught Auntie Terrible – whose name was also Tonic, for the dead husband to whom events and language would one day compel her to return – burying fairy tales under heaps of eggs cracked by her and scrambled by Americana, as if one could simply eat one's way back into that childhood condition of surprise endings in which one first admitted, with a bit of a shock, that one knew already what made that particular duckling just so ugly, and that one couldn't be less interested in finding it out all over again, thus preparing the inevitable cadence on one's tenth birthday and leaving oneself sixty, seventy, eighty years to improvise a cadenza.

"Do you know," asked Terrible, slyly shifting the subject away from questions of just who became that sort of subject anyhow, "whether the flesh-eating mutant who scrabbles in the crawlspace under your bedroom should be addressed as 'Mister,' 'Doctor,' 'Acting Director,' or 'Lady Thatcher?'"

So either the last animals were dying one by one into etiquette, or she had rooted out my hiding place, and was slicing me truffle thin one more time. But I was soft, and would not remain partitioned, being my own annunciatory golden shower washing language and décor together into a blur swirling in the strangely fertile drain. It was during one such episode, in fact, that I first realized how far into my private pleasures commerce and state power had crept, as my hand trembled over the light switch, undecided whether to make myself the spectator to my own minor and phantasmagoric perversion, or hang out in the scary and phantasmatic dark, understanding like a sudden electrical discharge that either option was a simple setting hardwired into the concept of "utility." This was when I made up my mind to look for eros in nominalizing the adjectival, so that a warm spray of yellow spilled beyond all boundaries as explanation for my constant schoolbus hardon, just as pink was no longer the color of lips, but the indistinct outline of the joys of orality itself. Sexy spectral siblings clustered round to translate this rush of sensation, this elevation, into a floorplan, in which my room had a door and a clock, and the world was peeping in, counting blinks against the darkness.

Of course, things had been different in my parents' time, who had learned just how far you have to dig into bedrock before you find yourself safe from the lights, those heavy glass artifacts full of smoke and congealed labor turned against itself, that fall from the ceiling and burrow into earth; and they would share this learning with me from time to time, more as a display of such curiosities as icemen, 45s, and other prehistories perfectly preserved for gawkers than as a principled attempt at instruction. Nonetheless that light, however ironically approached, however dimmed and deflated by self-lacerating humor, was the medium my sight clawed through as I taught myself to read in the dead and stony dark, against which experience, or rather the fact of having had an experience which was now irrevocably closed, like the smell of bougainvillea on the other side of cinderblock, had failed as yet to appear. My solution and, I would find after the fact, that of many of my contemporaries, was to cover every surface liberally with dents, nicks, scratches and pops in what I saw, perhaps not in so many words, as a humanizing deployment of the noise of post-humanity, sounds of friction slickening the slipstream, and mobilized as a filter against that "pure process," which was just then intensifying a series of effects such as the raw amplitude of building materials, so that blonde stone and green glass were driving crowds of people into the open *en masse* as if by some magnetic repulsion, appearing more and more, at least in the authorized histories, as ends in themselves.

"Nonetheless," said Grandma Violet, caught up in her Popular Front reveries, "it was a lovely tune when that dear old analog equipment took it up, a nice thick sound that I'm sure came from somewhere very specific. Maybe we should have guessed that something was not quite right when we realized the melody was always exactly an octave out of anyone's range, and all those screaming head voices started marring the beautiful realist symphony. Still, child, that whole 'tactical reversal of figure-ground relationships' song and dance is your hobbyhorse: I'd rather know there's something soft, small and fairly static under my aching rump. You might *believe* in that city *over there*, but I've seen it from a rather more intimate angle, surprised it at its bath and watched it chew its own toenails. It's nothing arbitrary – my body is not of the same kind as yours, since I belong over there, or over there belongs to me here, I forget which, but anyway it's not quite the same thing as a jar in Tennessee at the one extreme, or a random flasher accosting you in a crowd at the other. I have an acquaintance, Mr. Article, neither one thing nor another entirely, who is the objective damage wreaked by the city's having packed up and moved along down the road, leaving him a kind of indifferent yet fastidious cannibal freak always consigned to interstitial spaces not his own."

At this I started just a bit, recognizing the first cracks, or rather correspondences, between my private dressing room and the big, big stagey sky of the larger theater just outside, in which some characters took up watercolors while others had their heads chewed off. Usually these latter had been a bit too credulous in the face of Auntie Terrible's fairy tales, which were after all an accompaniment to consumption at which it did not pay to look too closely, the positions of predator and prey therein having been none too scrupulously separated, her acts of partition being, as we have seen, performed upon flesh rather too malleable to keep within the grid. Melisma, an undercompensated or volunteer sister to Americana, remained on hand during these episodes to enforce the supremacy of Terrible – whose body was the double exposure of her speech – as the one and only storyteller, whose tales demanded belief equal to our belief in a horizon of urbanity and urbanism, even as one had also to remain always unsure whether they narrated anything at all, the analogy being with our own continued suburban circuits unmodified by the emergence of a prospect. Melisma, as befit her name, was a mistress herself in the art of deferment. Americana, by contrast, arrived right on the beat, out of proportion to the delicate breads and pastries that were made to embody her huge ambition. It was in fact the aftermath of one of her deceptively insubstantial confections that had prompted Grandma Violet to recollect her youth, and the aftermath of that recollection was the thread or chain we all followed into a knotty and reptilian allegory of the body of digestion as that variously appeared against the ground of, or gave ground to, the body of the architectural sublime.

At least that was the explanation offered by Uncle Madame, appearing as was his habit from a place just out of sight, in the grand style of one whose sole commitment is to beauty, and who can make tropical fruits appear as if from just around a previously unsuspected corner in the studio he shares – in spirit if not in fact – with Mr. Article, consuming nothing and thoroughly enjoying it all, or so he said, though we'd often caught him gnawing at some perfumed flesh or other.

"Let my eyes be your private apparatus," he said to me from his near distance – in this, comparable to Mr. Bird-in-Your-Hand – as a counterargument to the insides Grandma Violet had left to me in trust, despite her being still alive, "and I will show you private views of public fact, such that even clockwork will astonish you."

So he took me to the movies, and I "told" the film at school to all my playmates, complete with sound effects around which my mouth strained to be a laser cannon or a natural disaster lending a credible urgency to the trading cards and other items of merchandise that followed in the wake of my recounting, surpassed it, and took it up in their own warm enumerated embrace. I imagined I was Carrie Fisher looking out, or listening in through the special radar devices mounted on the sides of my head, on my own boy-self breathing fast and gazing rapt upon my constant looking away, off-screen and up, in a gesture fairly saturated with ambivalent significance.

Uncle Madame was not above renting the occasional semi-pornographic art film – heavily edited by Blockbuster, at all events – for viewing with the assembled family, simply for the sake of making Grandma Violet go as delicate and colorful as her namesake in a paroxysm of tolerance. This had backfired once before, with a video in which a boy sharing my name had featured heavily, and Madame had kept up a near-constant stream of observations that I "made a better door than a window," with consequent demands that I reposition myself relative to the screen, as a means of keeping me from observing too closely the mechanics of exchange whereby a succession of satin-finished and vaguely-gendered ladies presented the boy, his silken collar addled with containing his sweaty neck and dry throat both, with an inexhaustible stream of velvety tropical fruits, kept just out of sight until the decisive moment in what seemed impossibly small purses to have accumulated such a trove of neocolonial treasure.

"There used to be a name for people like us," Madame muttered sadly, as he realized the depth of his error simultaneously with the liquidation of any standard by which to judge it as such. "I'm sorry to have placed the boy on view, not because it makes a whore of him, but because thus uncompensated it will never make a line on his résumé. I only thought to do a good turn for the boys and girls on screen, to pen an episode for the special guest stars who look so witty but so blue without our clumsy company."

THE SCRIPT-DOCTOR, OR, THUMB-WRESTLING IN THE REALM OF NECESSITY

My rule was, you gotta earn a direct address.
–John Cusack

"Skeletal necessity" was too often
poor substitute for true inversion,
pressed diligence, original in absence of a difference.
One choice invests "would have" in time.

"What imbeciles we headstand artists are!"[3]
Look harder now for deader giveaways:

—If a snake is what I spied, proprietary laze upon the lawn.
—If a premonition is what coughed hard labor in your eye

and poverty stretched out beyond the eight-hour day.
These things we had to learn and then forget. Now who

but cardinals wait the time-clock temporalities of spring?

[3] So there were footlights, or headlights, and my spectator's pleasure was quickly consumed in utility and blind curves above the speed limit. *And with my eyes fastened on that inconceivable image, I strove from morning to night to overcome the barriers which my family were putting in my way.* Which was, of course, to emphasize the "work" in this "opera," bracketing the question of improvisation to concentrate on heavy lifting, regulated by exhaustive time-and-motion studies, stroboscopic film strips catechizing the body into a correctness of posture and a rote, mechanical exhaustion only in whose wake could the proper sort of prurient tingling be felt.

Moreover, as the busted head lingers / as the customary life hangs
frames alone around an empty focal length…

…you'll need to add vaccine to what your eyes are flush with,
call out orders to the ants that rush and skitter
in the corners of your vision, form them into columns.
Of what ballots then is that inheritance made, the bolder
as each new murder is determined as the world?

High on the pronunciation,
we'll issue clearance for the shunt

of former need to future
reservoirs, *drink to me only*
with hydraulic power to intend

my face at least —
You order me
to go, rapt
and wrapped for takeout.

No, the opposite tack is much more clear:
rage with a pair of household implements
calmly makes refusal, quietly takes up slack

"*My uncle, who hated lending people books,*" I imagined myself eulogizing in silent ventriloquist's advance how a past literature might remember him, and thus have provided a foundation for the bind in which we now found ourselves. It didn't yet occur to me to wonder how the rest of the family simply sat absorbing this, my brain being not yet sufficiently altered by the chemical stream whose floodgates I first would come to open in the days to follow. In not breaking with this blankness to look at Uncle Madame at that moment, I opened instead a different set of gates, full of brief flickers of personages even less substantial than Melisma, and guilt blossomed all around us like a cathedral decked with azaleas and microwave relay towers, in which every person had assistants and assistants of assistants, and so on in a non-linear and interlocking progression or web whose nodal places changed on average every two years, giving the illusion of an infinity of less than satisfactory arrangements for dancing, dusting and dining. There was juice pressed out in all that slow and purposeless trampling, but no one cared anymore, least of all the actresses, who from that moment on carried instruments sharper and more crooked in exact proportion to their own inability to use them except to hand them on, thus substituting "utility" for "fruit" in what had been until then a lovely and imperfectly glimpsed, clandestine sort of marketing.

"Oh yeah, an integral for your loss of integrity, dig," drawled Mr. Bird-in-Your-Hand, just then recalling himself from a haze of sweet, illegal smoke, "having as upper limit function, having as lower limit function, itself decidedly a function, and entirely within the limits of the functional."

The "skeletal bulge" invests an imaginary
frame alone. A vaccine added
to ants' column by ballot, the bugled
shunt of friction, slack.

Silk in colorful addition accumulates
the lapse of time as stacked
palettes, lulled askew the
collated page by lien of sight.

Notation's ***I***, its threnody a web,
troubles one by tone of brick.
Money praises color, units cut
by drag of frame "trifles"

caught on fulcrums turned into. Face
their halting light, smothered face.

The impassive, unsympathetic, sublime face of true goodness.

Against this accusation of a stupid innocence[see "Correspondence," note 3], this plot to file down the very plate from which the image of my face was pressed, overnight express mail dropped with a sudden boom to the floor, filthy with dried blood and buzzing flies: a nearly pastoral, summery kind of ambience. The heat haze thickened all at once, spiritualizing the landscape in a big hurry. Looking out the window, I only just had time to make out the squat bulk of the abandoned mall, in which bustling crowds of people figured only as an excision, providing me with a real grip on symbolic logic. There was a storybook inside, still available for discount purchase, through which I might come to have the kind of body that comes into a story, if only I could get into the building before the crews arrived to convert it to an indoor flea market. There was a riot of color through which to filter all the other uprisings – so-called riots – then in preparation, even if we were not ourselves owners or even gardeners of the flame-like flowers that brightened and obscured our view.

"That spirit in which we believe is simply the mobility of liberalized exchange and currency speculation," I muttered to no one in particular, my very first public utterance already compromised by the self-canceling hum of the literary, "*for even if we have the sensation of being always enveloped in, surrounded by our own soul, still it does not seem a fixed and immovable prison.*"

That no one listened was not surprising since, truth be told, I had cribbed my pronouncement in full from the work-in-progress of a skywriter wheeling overhead in an antique biplane, and the hour of talk I had missed in thus attending to white characters inscribing themselves softly on the hard blue had carried the conversation far beyond me, so that when I returned from my plagiarist's study, it was to a room radically emptied of my presence, and in place of time served I was debited with space taken. In this sense the day, as all days in those days, became a nearly perfect circle around me, interrupted only by the frequent flyovers of the military jets from an Air Force base on the other end of town, a sound I would have ignored had it not been taken up in imitation by the little girl next door, her tongue making of her mouth a turbine to stick against one of the diamond apertures in the chain-link fence and suck a bit of wind out of my sails. Americana feared both girls and armaments in equal measure, hearing their noises as the roar of two distinct sorts of maw that might swallow her equally whole, operating at such a level of scale as to have no need of the head-gnawing with which Mr. Article at least engaged one's body in a sort of negotiation before consuming the better part of it. Speculation and argument ground to a halt, however, as the girl's father, who was the foreman of our landscape crew, and the real topiary force behind the bright and ideological garden in which even we mulch-spreaders and deliverywomen were allowed to live as part of the treaty of exurban compromise, arrived on the scene just behind the little terrorist and, pointillized by the fence into a neutralized aesthetic sensation, snatched her up, stuffing her deep into his hip pocket and in the same motion making a mock-apologetic bow to Americana that took in the rest of us as well, saying graciously with an authoritative lack of accent, "I advise you to disavow what you have seen this day, if that is what you most desire."

At a savage pinch from Mr. Bird-in-Your-Hand, I was made to watch as my beautiful and balanced skywritten treatise unraveled into air pollution, which was not forgetting but having pressed into my wet and – thanks to constant gnawing – exposed brain a more permanent and subtler sort of lesson, not out of agreement with my original intent, but occupying perhaps a more habitable space, and saving me from the life of libidinal complicity held out in the little bit of loam – or was that a stack of densely scribbled pages? – piled in the foreman's open palm. It was this minor literature I took with me to school next day, sharing it as I had my movie at that earlier recess, and it earned me the half-mocking nickname "Mean Little Daddy" from the other smudge-faced miscreants, as we picked our noses and expatiated on the cultural topics of the day, or at least the day before. That we lived, by default of income, in one of the few truly integrated neighborhoods in the city, a neighborhood, moreover, whose characteristic architectural operation was the surgical removal of anything at all characteristic[see "Correspondence," note 4], simply meant that we grew more than usually adept at teasing out the narratives and categories of each other's origins, assigning them to each other in much the same position as the name of an employer might occupy on an identification tag worn at a business convention by those who lived, more or less, in other neighborhoods entirely. When it grew too hard, too hurtful, we took pills and assumed that we were aliens.

* * *

"I think my family imagines the overripe fruits, gone soft almost to the point of putrefaction and still suggestively tumescent, that you mash so gleefully into my dull and unexpectant face, as props in a ritual too otherworldly to script as the simple practice of a man who comes to dinner and then leaves," I was saying to my friend Vatic Bill, whose gestures were sudden, large, and sited against the most distant amphetamine horizon called everyday life. "Your proximity, even at this remove, makes them imagine that they might be themselves exaggerated, too Hollywood by half, as if they were actresses reprising a sexually insatiable and half-concealed past in an affair with history on the other side of the undeveloped acre of woods, its derelict couches, TV sets and discarded tires mulching down into a common ground no one treads except as a shortcut to the Handy Mart with its attached discount beauty college."

Vatic Bill had introduced me, earlier that day, to the habit of not returning to the point, crouching interminably in the open mouth of the dry storm drain where it hung over the concrete culvert, ready to empty the streets into the water management district's magnificent and deep new contingency plan should the present drought ever end, and experiencing this recess as precisely that, a hidden alcove or "inner prelude" within the public works among which we lived, and where the possibility of a beauty radical because unsuspected disclosed itself, even as we bent double to fit ourselves within it, smoking cigarette after cigarette and littering the stretch of concrete pipe nearest the aperture with the debris that gave evidence of our belief in its status as a kind of fictional parenthesis, a bit of the local engaged over and over as a makeshift travelogue mediated by a professionally detached and competent voice-over. Things continued to happen: they had been happening in fact for a few minutes before we arrived on the scene, and sometimes we took this as confirmation of our prescience, as if the fragmentary dreams of hidden crawlspaces, of whole underground cities in miniature, like the one rumored to exist under the emergent subdivision a mile down the road, and whose obverse was the chthonic sociology that produced our minor monsters like Mr. Article, were citations of something like totality, as such and in full, the universal returning in the mode of a repeated "according to..." at the head of all our assertions and gestures of escape into perverse particulars. This sensation was further confirmed when I tried to puzzle out a way to thank Vatic Bill adequately for an introduction thus indispensable, and then, touching him tangentially at an odd moment already after that first awkward act, blurted in an almost accusatory tone, "*My feverish and unsatisfactory attempts were themselves a token of love, a love which brought me no pleasure but was none the less profound.*"

It turned out, of course, that all this was only impatience for that critical self-reflection I imagined to be the shape and circumstance of adulthood, and that would institute a baseline of constant low-grade regret into all my doings, as bodies and their interconnections became both more desperately scarce and more easily substitutable, dry goods overlapping haphazardly on pallets all askew in the dust and clutter of a disused warehouse for memorial stones – or so the various adults told me later that afternoon when I returned home and painted them a picture of what had transpired in our circumscribed and subterranean world.

"Hey, little man," Mr. Bird-in-Your-Hand whispered confidentially as this ideological apparatus began to break up around us and prepare for dinner, "I hang out sometimes with the guy whose company built that very pipe – well, at least they dug the trench – and I can probably get you a job, or anyway a nametag from one of their jobsites, if you'd rather just fake it."

Mr. Bird-in-Your-Hand was always going on about how much better it was just to fake it, which put in him in the curious position of being the only sincere adult I knew, always resolving the difficulty of this function by disappearing into spatio-temporal anomalies in some minor narrative or other in order once again to escape into sardonic and self-defeating presence in a quantum flash one knew was happening only after it had passed, as the sonic booms from the military jets often rattled our windows only when the planes themselves were well on their way to yet another dangerous and demonstrative flirtation with Cuban airspace.

"His name is Rudder-in-Law," Mr. Bird-in-Your-Hand droned on, insinuating the contractor into the echo chamber of my geopolitical reverie, "and he wanders around with his eyes out in front and the stone cold power he's out to admire not far behind."

[Tutti]: After So Light…

[Mr. Bird-in-Your-Hand]:

…demolish to compose the girder as such
                    literal: your hand and then shut up.
and poverty was past the working-time
                should emerge face from its haircut
recycling unpaid overtime. Not red-handed:
palms flush when I touch
your unused universal alphabet
My face at least —
a household implement

Had learned and forgot 100
foot gypsum stack in 300 foot hole.
In lapse of a long think never fear
            who but
the photograph helps you feel the loss.
    voluntary fruit of force
you order me, correspond

to nothing — tip-in to tip-off

calmly and quietly slackens
Dubious transmission of
voice says "dead weight" means anyone
To function! — and not know I speak the truth:
appalling, pulling at your sentiments. Come closer.
a pistol followed such destruction she says
you trivialize in use.

Had learned and forgot
diagonal to prevent such warping
—swallow move the mouth

And suddenly there was that girl again, making little fart noises with her lips on the heel of her palm as if commenting on Mr. Bird-in-Your-Hand's aria from just out of sight behind the kumquat tree in the next backyard, so I knew that becoming what I have in fact become might fall into thousands of stony pieces were she once to emerge as the sort of person with whom I might thus walk around and see the sights I looked for. My sense of the world was as a club, at which she might get my name placed on the guest list, were I ever to get beyond the environs of ditch, scrub woods and citrus with my eyes as open as the aperture claimed for his doubtless half-fictional friend by Mr. Bird-in-Your-Hand. Just then Auntie Terrible – who had names to spare – began to get a sense of the drift of our clandestine conversation, and hastily arranged for one of the impromptu slide shows whereby she domesticated the erotic shock of places as remote, baroque and ornamental as the abandoned drive-in theater just off the southbound lanes of the interstate that ran through town, exhausting their signal difference through repetition in a series of badly cropped and still too-symmetrical photographic and theatrical nominalisms.

"Besides," sniffed Americana, imputing by the gesture a grainy dryness to the light rain that attended our every excursion to the inside of the house, "it's opposite day. Again."

Something was trying to get through.

While the phone rang and rang, Grandma Violet made herself busy answering other appliances altogether, failing on principle to distinguish between the frequencies that announced the coming of a person and the objects that were marks on the world left in his or her wake. Melisma, ornamenting Auntie Terrible's ironic commentary on these developments with a fine filigree of philology loud and effusive enough to drown out Americana's sober protestations, heard the word "wake" and promptly kicked off during gender reassignment surgery, returning to live among us as a dead man before we realized he'd been gone, so that everything we said to him thenceforth was in fact an item for a eulogy and a double entendre playing over the surfaces of his picturesquely scarred body.

Somewhere between, laminate states somewhere
other. Midwest "maybe" addles plain,
or wit unadorned. Simplicities
skew under brick. Strangle licking

into phase, quiet conflict, predict,
devote instruction, prepare to reach.
At grasp or gain, again.
Home style, see, a zone without

position, not air, legitimate names.
O violets! No credible detail
vaulted tiny flames' wet prehensile
genital nationalism. Person you

rub, not carton, yet are taxed,
rife with serial severance.

"Whenever he saw a man in a town whose face he didn't like, he would massacre the entire population," black out the relevant passages in the archive, and pour water on the photographs.

All of us had our various and variously culpable methods for trying to make the landscape cohere – Pop's driving on the sidewalk for example, which bode well to return some of the possibilities of *flânerie* to the minimal densities of a vast edge-of-town not so much sliced into districts as spread like imitation-mayonnaise "salad dressing" over a hypothetical and largely unoccupied grid already thus sliced, and in which the desire to be consubstantial with the body and perspective one might have, say, from the other side of the retention pond, made perfect sense as a speculative maquette of the religious feeling for icons, or the social feeling for the face in the crowd.

"But it's taxpayers like your poor aunt who take the biggest hit on this uninsured swerving through the putative plasticity of sex and cities," wailed Americana, cradling the head of Auntie Terrible – whose name was the one word in her life certain not to change – and rubbing two pennies together before throwing them by way of sympathetic magic into the pot we used to piss in before we plumbed the depths of our enmeshment in the grid. Someone, probably the long-neglected Auntie Trifle, answered the phone, plugging the exposed ear with a finger so as not to hear Americana fairly shrieking, "Invite whoever it is to the opera we're staging – or the zoning hearing we're watching on cable, I forget which, but anyway we have lots to spare."

Whoever was on the phone – it might have been Trifle, but there are some connections without regard to persons, after all – launched a volley of exasperated shushing our way, the force of which spun Americana around a full 180 degrees, coming to rest in a finicky inspection of the blood under Melisma's fingernails, giving Melisma by the same gesture the perfect opportunity to survey his mirror image in Americana's glassy stare. It was at this moment of supreme stillness that I noticed Auntie Terrible, whom I'd always taken to be immobile herself, running in place in such a subtle manner that, were the staging less perfectly balanced, it might have continued to escape my attention entirely, energetic though it was in its own self-effacing way: she was struggling just to keep up. Listening closely, I caught her muttering under her breath while the world of telephones, debt, traffic citations, and mutual recriminations went on around her: "Stay awake, stay awake, dead men come to marry you and the assistants can do multiplication with their fingers."

This whole sequence of events, from the phone call to the talisman against structure in dominance, jumped out of the course of our days just a little, repeating on a somewhat slower pulse and taking on the appearance of the unanticipated so essential for epics and block party nationalisms, the dangers of their deformation countered by the neutralizing effect of their reduction in scale, so that the premature finish to what the microwave was doing became an end in itself, and time took on a ready-made gratuitousness of flaky crusts and high-fat fillings, right on schedule in its absolute surprise. When I call this surprise, I mean something fundamentally distinct from *the surprise of a "barbarian" (for so we termed everyone who was not acquainted with Saturday's special customs),* and more along the lines of the annual exchange of secular but still fully obligatory Christmas gifts – more obligatory to the individuals caught in their orbit, in fact, for all that they were unattached to the obligations of an authorizing context.

"Auntie Terrible," we pleaded, trying to extend for as long as possible the moment before the inevitable burning of tongues, lips and palates on overheated, overprocessed cheeses, "lie to us about what Uncle Tonic said after he stopped speaking, but before the story ended."

"Well," she began as always, "at the time his name was not Tonic, but Twenty Times Yes, for the effusion of prenuptial humanity in him when he was still composed of gestures rather than fates, and in fact he announced herself to me in such a gesture, *a swift and thoughtless movement of the head, with a provocative glance from her contracted pupils, by a young girl in white, insouciant and vivacious.* If you've ever had a girl look at you through her pupils alone, precise and analytical as close-miked single instruments in a cavernous concert hall, then you know how badly you want at that moment to pull back for some perspective, even give yourself over to boredom and loss of desire in order to preserve your experience of what you still, in your premodern phase, call 'the whole.' Enough of this humanist nostalgia and the architecture did in fact intervene, and quite decisively, until *whenever she spoke, she heard her own words with the ears of those to whom she addressed them*, and this made of her a wholly different sort of person, so that hearing the too-deep pitch of her own voice while flirting with me – as a car flirts with the shoulder, perhaps – *immediately it would seem as though a more sensitive sister within her had blushed at this thoughtless, schoolboyish utterance,* and from the self-corrective excess of girlishness with which she sought to address this imbalance, *he* emerged, Tonic in all but name, with nineteen affirmations still to get through before the name change, catalog page after catalog page redressing his sumptuary crime, which was the really unbearable, fatal part, like watching a Slinky walk down a flight of stairs, inevitable yet fraught with a degree of positivist risk."

Right on cue the phone went dead, as a shaft of pale moonlight angled in through the window, reminding us that dinner had grown cold and some unheralded disaster had swept away another bit of infrastructure. Were car alarms cycling through their repertoire of four or five rhythmic patterns, or were we supplying those clarion calls to nervous inaction from memory, and *was the ground then moving forward under our feet*? Auntie Terrible theorized an end to the clocks more dramatic than a running down, and waited for a branch to fall on the new roof as if that might confirm her in her minor messianism. She was always the one who remained to tell the tale, the hostage recently released into a press conference, spending precious airtime freely in remembrance of those who, one assumed, must have died to bring her face to face with those cameras. In the clearing opened by the fall of our house into its prospective deliquescence, she stood firm to hate Americana with a purity undiluted by any admission into dialogue, having scripted the battle in advance and determined that she herself would lose, and badly, while the detestable faith and industry of her assistant would rise triumphant on her still-protesting bones, which might yet have occasion to jog in place for the sake of appearances.

Of course, the only reason she kept this specter of Hegel standing before her, chiding her with a constant "I told you so," was that, being almost entirely reflective, it served her as a paranoid rear-view mirror, a prosthetic set of eyes in the back of her head opening a line of sight on the floorplan. This helped her to maintain a certain degree of coordination, since most of her conscious motor control was given over to an elaborate mime she'd begun after reading a painstakingly researched biography of Buddy Nation, the cartoonishly masculine automaton entrepreneur who lived alone in a Plexiglas house with the lights always on – really, less a biography than a sheaf of deliberately leaked surveillance documents, but none of us had the heart to tell her that – and which had now progressed to such a rare degree of refinement that her every gesture was socketed into belief so firm as to approximate the positive knowledge that it corresponded precisely in contour, intensity and timing with a fraternal posture performed by the unwitting mechanical subject of her documentary scrutiny. During the same time, we had gone from eating our meals on newspaper to eating newspaper, and so had no room to speak. Nonetheless, the suspicion was growing that someone on the other side of town found our whole diminished tableau just too cute for words, and wondered from afar at the lifelike movements of our precious little limbs when not watching us from close beneath the house, smacking his slack lips as his body spread out just below the surface of the neighborhood, minor tangent at a major scale, like a huge and barely coherent rhizome growing as a net under the loam of Michigan. So there was this sense, in the wake of Auntie Terrible's story, its sublation, and the hastily gulped, cold supper, and though Florida was far enough from the upper Midwest that the last instance occurred to us always in the form of an exemption to be claimed, that we were already eating again, and that some labor of consumption was always being prepared and carried out for us by the ground we thought to house ourselves above, just as the hard, blue sky was housed above the floor of ozone haze we saw when we looked up, wishing it were possible simply to stare into the sun for a long, long time by way of feeding our insatiable appetites both for mortification and for appetite itself. Once understood thus, the whole pattern cast my episodes in the shower, previously understood to be caught in the contradictory grip of private impulse and administrative reason, in the far more quotidian and public light of what I remained naive enough to call civil society, making it clear that what I sought was the trace of recent history in the varying smells of urine atomizing and diffusing through the superheated steam that fogged the windows, mirroring the exhalations of the streets themselves that evening as the rain let up, just in time for us to drag the TV out into the yard and gather round to watch a bit of wrestling.

"Kill the motherfucker!" Americana screamed, getting things going as usual, while we comforted ourselves by pretending, again as usual, that her outburst was directed at the 700 pounds of mingled, spandex-wrapped flesh onscreen, and not at some unseen presence on the other side of the neighbor's wooden privacy fence, over which she was just that moment peering. *While conscious of her cruelty, we appreciated her services* to our own slow-emerging literacy, or latency, it was never really clear which. One thing we did know: the papers were full of whatever it was, and we wolfed them down with relish every night, converting Americana's murderous encouragements of the night before into the sympathetic vibration of breaking stories in the bowels of the present moment. We managed to digest a fair amount of world during evenings like that one, enough in fact to save me, in any other than exceptional circumstances – but do we not always live under exceptional circumstances? – from having to venture out into its freshly bathed and steaming streets, past the huge banana spider that hung in the bushes just outside our front door.

As night fell, Mr. Article – for it was he who was, in fact, the dimly sensed and hungry fungal presence beneath the floor joists – stood up and stretched, shaking us off into the fine gray dirt and sparse grass, from where we watched his flesh coagulate, collect, and comically wobble downward as his frame grew more erect. Yawning and reaching ever higher, his hand brushed the electrical and telephone wires, and this burst of information rendered him suddenly and ecstatically complacent, so that we knew in a blinding flash of voltage there would be no perforated skulls, no leakage of thought into the greedy sinkholes sucking at our foundations as we slept tonight.

"Come and let me breathe from the far country of your adolescence the scent of those spring flowers among which I also used to wander many years

ago," he boomed, channeling and misdirecting crank calls our way with all the heavy breathing filtered from the soundtrack. We chattered excitedly among ourselves, suddenly animated and galvanized by this language of fragments, absolutely revealing and every bit as inadmissible, so that what we knew each other to be from local news reports in which one of us or our neighbors always seemed to figure as a cautionary tale, shot in the dark with infrared cameras, in no way corresponded to the shape of limbs we patted reassuringly in the blind aftermath of the power surge.

"Excuse me, sir," I petitioned the unexpectedly amiable cannibal giant looming against the sodium vapor lamps that were just that moment fizzing on, "is there a war going on?"

"Fee fi fo fum," he recited, loud as hell but bored to tears, "that information is only available to our subscribers." [see "Correspondence," note 7]

Then, sneaking a quick look over his shoulder before loosening his grip on the spread of other people's news, he leaned in close to intone confidentially in the tones of polite menace, effectively breaking the electrical promise of détente his earlier appearance had made, "In that land to which I voyage through your daddy's mind there are rough boys who will punch you hard in the sternum until you stop breathing and live as pure aspiration. Thou shalt not follow me across the turf of that country club, golf ball thief though you be."

The division which continues to radiate outward from his fiat, like that evening's protracted and pollution-enhanced sunset, in my memory ever refining its rays in closer approximation to an ideal either sublime or wholly insubstantial, is like having a job which is my own and no other's, which is in turn like having a unique piece of money, which is finally like not having money at all, but a dark and musty store of something rich that can't be spent in the telling, a coin collection kept at the back of the underwear drawer with the disused socks, outgrown briefs and pornographic magazines found bundled into plastic trash bags in the woods. I remember now that we had neighbors living close on either side, and that I had brothers and a sister, and have them still, though I seem unable to mention them other than in passing, passing by a way that is not named in the habitual choice of front door or back. Perhaps that was what the right side of my body was always up to, while I wrote these things down on the left so as not to forget. I notice again in this connection that people whose movements are restricted have the shape of trees, and remember that this appearance was our habit in the company of strangers, so as to keep our forested interiors out of sight, and wonder whether my siblings have simply remained as entries in that protected column, clandestine and confidential, a certain value and quantity of small-parcel landscaping in this connection of the recollected real estate of low-cost home ownership to the several yards of county easement buffering front yard and bus stop.

* * *

Certain places persist in remaining surrounded by the vassals of their own especial sovereignty. A red plastic lunchbox, for example, represents a particularly exposed position, superimposed in all its collectible rarity on the usual succession of free lunch vouchers presented with half-averted eyes at the end of the ridged metal tray-slide, a unique instance more akin to forgetting than remembering and thus gathering as its threads the unlikeliest elements of the poured concrete apron back of the cafeteria – dead lizard in the seam between two slabs, bit of milkweed pollen blown in over the fence, seagull stabbing at the butt end of a Twinkie – into a synthetic pastoral on the far side of construction, a country of sorts whose constitution is equal parts reaction, in both the chemical and political senses, and radical anticipation of a work to which the hand has not yet turned. Or so I thought to myself at lunch on the day following Mr. Article's joining himself, by way of drawing distinctions, to the inchoate nominatives of my own existence, and the high sun, just as eternal and immovable as its horizon-hugging counterpart of the evening before – really another body in another space and time entirely – glared whitely off the sidewalks and cinderblock buildings, gently battering all the other children into slow, early afternoon stupidity and silence, in which this dull thud of thought, of which I accounted myself the only auditor, could better resonate and spread its influence, a web of unseen forces whose integrity was confirmed by the total lack of resistance in the visible society arrayed before my squinting eyes.

"If only," I thought to myself, "I could liberate in similar fashion the human sculpture held in abeyance within the block of this moment, lift the essential oils and odors of these grubby fingernails and sweaty crotches that surround me in all their ephemerality into permanence and providential bloom, then the sequel to this moment need not come in the fashion I suspect it must."

They would say "what do we do, we've never played these things before." He said "that's the point. You're playing from ignorance – it's an exercise in ignorance. We're going to play what you don't know and what you don't know is huge."

— John Szwed on the recording of Sun Ra's Strange Strings

A whole new category of police will have to be invented.
– Erica Hunt

But my inexperienced and unsubtle nose gave me away, and the rough boys rose and punched me in the chest until I fell on the baking concrete[4], where they left me gasping as they stalked off, telling the racist jokes their fathers had taught them by way of casual conversation, and marking in this way the bad faith they would require before admitting me to their aromatic huddle.

[4] [Eds.] The reader is referred to the following text, produced collaboratively by the editors and originally published in broadside form as "A Commentary on His Tiny Body."

Memories' ilk out there. People notice
the host intelligence slaved to the switch
obvious physical loss
marks the viewer's face: sebaceous semiosis
even those who disturbed the contrast
wet where kissed, slickly unwritten
scale — maybe the mirror as

new fingers
Horizon tightens till it penetrates, rewrites my insides.
assure itself complete before
Click here: "information" implodes.
Line discords *that* to every *there.*
mail-order x-ray and no glasses
reported speech battered — immediate purpose crystal
fear and non-language

about the city called flies
to the object which does not listen
pulled with a hand
your voice as PacBell nibbles at my ear.
translation forward
not salvation but redemption — cash value
or mended perfection by hydraulic
inflammation of the soft tissues
dancer, to close

through love upon their death, such
aerial photographs of the wilderness grid
more often (than) imagined

We suppose we must be trying now to confess a certain sexy glaze over the surfaces of these visions, rising precisely from the pink and repugnant uselessness of those boys, and our conviction that the narrator might be the one to bring them back from that condition into an understanding of the duplicity with which mere exchange has blinded us all and glazed our sight[see "Correspondence," note 8]. Across the scrim of refracted light cast off by the sudden downpour in which he collected himself and stood again, shaken but self-possessed, stood a girl, *a vision that appealed not to his eyes only but required a deeper kind of perception. If her face had not been so brown, he should not have been, as he was, so especially enamored of its imagined white-*

ness, nor come to the sudden realization that we were neighbors, and that in consequence my untroubled and jovial sense of how such things were negotiated without incident anywhere other than the cafeteria, a friendly notation of minor differences in tint that might form the basis for a nickname if not much else, was composed almost entirely from the raw materials of bad habit, worse politics, and eyes deliberately glassy with the penetrating under-frequencies of educational television, in which the representative function of Maria, Gordon and Mr. Hooper with respect to the racial typologies of Sesame Street was so visible as to disappear altogether.

"You don't necessarily mean shit to me," she smiled my way, "so hi, bye, and maybe I'll see you around."

With that she hid her face from me but did not weep – the rain was taking care of that, all on me – and she seemed instead to take on a power out of all proportion to her age, exceeding even the encompassing horizon of maternal directive, against which I half-hoped for an unlikely accident in which her own mother, arriving to pick her up after school, would find me unaccountably standing in the traffic circle in front, rather than, as was my usual practice, waiting for my bus in back and, unable to process the surprise my sudden apparition might thus occasion, would run me down with her heavy mid-60's Impala, so that, disentangling myself from the hard angles of the undercarriage and emerging unscathed, I might demonstrate that, while far from being a similarly irresistible force, I was at least a minor league species of immovable object. Letting the compensatory fantasy play itself out, I imagined myself as both a blockage in their toilet and the plumber who arrived to remove it, opening up the elbow joint between my own quotidian disarray and a transcendent state of harmony and sympathy, in which I still at that time believed, as if this conflation of the promised flush of my own first assisted orgasm, still just over the horizon, and the also fast-approaching group hug globalisms of the early 1980's, were the one true goal of a politics of everyday life in which it was even possible to forget, or sublimate, for just a moment, the stubborn fact that the girl's father was already himself, in fact, a plumber of some repute.

Arriving at home as usual despite these flights of fancy, I asked Auntie Terrible about the proper response to such a difficult situation, and received for my trouble a troubling story about the sandspurs she stuffed into her socks, a few more each day, until the gradual diminution of her capacity for social action reached its zero point, but with the conviction that this stasis represented retirement and expensive relaxation, rather than self-mutilation sold cheap enough to devalue even the steps others might take to maintain the sidewalks along the public ways. I took to standing at the edge of our yard, directly under the nearest window of the girl's house, still unsure whether this was the side on which a chain-link fence and kumquat tree separated me from her, suspecting I had seen that armature somewhere else entirely, or else had seen it here in a peculiarly filtered early evening vision in which light bent around the heavy moisture in the air, but leaving that question aside, would turn my face back to my own wall and call to the family so that, running out to see what was the matter, they would be forced in looking at me to look also into the window of all the objective life I had lost in coming to have the subjective image of my relation to it, feeding my hope that in this arrangement there might be a kind of triangulation through their eyes whose indirectness would allow me to glimpse again the solid contour of a girl whose name, Dolores, was the least allegorical of all simply because I hadn't made it up, a hope I maintained even if its complicated stage direction meant perverting my own parents, turning them toward a fetishism of the look of things that caught up whole people, whole histories, whole peoples, in its process.

That first evening, though, leaning jauntily against the orange tree which came up to the fence on our side, as if daring them to remark on the nonchalance of my posture as it neared puberty, I managed to rub the powdery green fungus growing on the bark all over the back of my one and only good shirt, worn for the school picture which had been shot earlier that day, and this indeed provoked a response from Mom. Geometry suddenly assumed a paramount importance, while all my other speculations grew ethereal as the phantoms of sociological knowledge upon whose field Pop sparred with the spectral person of my emergent stepfather. There were whole volumes of the stuff to be read, and no point waiting since, ethereal or not, it wasn't likely to go away, so they started me off that night, leaving my lights on and tucking me in under a stack of books, giving me perhaps too much credit for the best of intentions and too little breathing space. *Physical passion, so unjustly decried, compels its victims to display every vestige that is in them of kindness and self-abnegation, to such an extent that they shine resplendent in the eyes of their immediate entourage.*

"You begin to get a feel for the narrowness of the margin to which they relegate these questions," whispered Twenty Times Yes, who appeared to me that evening, left out of all narratives in which her presence would have been an anomaly, and stuffed without ceremony into a toy chest, out of which she heaved herself into the night light's orange circle on the floor. "*The facts of life do not penetrate to the sphere in which our beliefs are cherished.* But things can and do change: only last night I was powerless, unable to make an appearance at your Auntie Terrible's bedside, and now I have the full strength of my own principled unwillingness to do so."

So saying, she pulled herself up to her full diminutive height, reminding me in this gesture that she persisted in a state of accelerated arrested adolescence, having been rationalized as simply one more item of off-the-rack clothing with which Auntie Terrible had dressed herself for the appearance she was now in continual process of making, and recalling as well, by her distant look through the darkened window that could not have been at all revealing to any but one such as she, who was accustomed to reading the minutest marks for signs of intent, that a condition was held to be imminent by all save her, in which *the brightness though not the luminosity would be expunged from a landscape in which all life appeared to be suspended. A tradition at once ancient and direct, unbroken, oral, distorted, unrecognizable, and alive.* Whereas Twenty Times Yes militated against participation, being built somewhat more like the motion of an actress crossing the background of a scene in a film, a blur of indeterminate color values one by one deeding all their foregone precision to the parcel of scenery they managed to displace, than the well-mapped planes making up the image of the same actress in a photograph.

"Word travels to you quickly enough," she laughed, "but your reflexes are damped down, weighty and syrupy, as if you live in some experimental underwater colony. You've imagined the problem between you and this girl so vividly that, congratulations, it really is a problem now, oh and by the way, I think your aunt just finished drowning in another room."

Running out the door and down the hall at this expected but still unwelcome news, I came face to face with Americana, standing vigil over the grimly occupied bathtub, blocking my view and howling out a big angry love song that read like ritual violence.

"Oh, if only she had burned, like bad bread or cheap fuel," she wailed, then and there helping me to decide that, for all time to come, I would hate little bastards like me, who could not be bothered to take the sentiment behind the lyrics seriously enough to learn the tune. Yes, this confusion and lack of rigor would permit me to stroll pleasantly through a landscape in which my immediate urge was to decapitate everything else that came up to my height for the crime of sharing in those very qualities my stroll was compact of, an itch I would never learn to stop scratching at the least sign of its renewed inflammation, sloughing bark from trees and flesh from limbs in equal measure as I rubbed, panting with gleeful vigor, against whatever stood to hand. *And if I had just been thinking of my parents with affection, and forming resolutions of the kind most calculated to please them, they would have been using the same interval of time to discover some misdeed that I had already forgotten, and would begin to scold me severely as I was about to fling myself into their arms.* Therefore I wished constantly in the wake of Auntie Terrible's departure for parts unknown that I lived in a landscape better able to concretize my impulses to have done with the impermanent connections, the permanently low signal-to-noise ratio, of the wet and tepid mechanics articulating one human body to another, that such a landscape – say, New Jersey – would provide me with the materials I would need, the grit-enriched snow out of which I could sculpt the other bodies to whose cold, abrasive presence I would make love, over whose hot and melting absence I would mourn, in a pattern seasonal and cyclical, not so irremediable, not so dredged and filled once and for all. *And between the earth and its creatures I made no distinction,* which, had word leaked out, might have foreclosed permanently the possibility of a continuance with Dolores and the other kids, poised ambivalently as it was between worshipping the ground they walked upon, and transmuting their bones to the sticks and stones whose names would hurt my tongue less in the saying. *One does not think of oneself, but only of escaping from oneself,* and against a landscape whose every feature was marked with terror – people could show you, for example, the oak tree whose largest limb had been part of the apparatus of the county's last acknowledged lynching, only some ten years past, or the wall, stained slightly down near the ground where the color of bricks went from red to brown, against which skinheads had broken open a kid's head after a concert the summer before, or the interstate overpass, whose protective cover had become a permanent neighborhood everyone pointedly refused to acknowledge on their way to work – this turns out to be the most stupidly subjective detour of all.

MS. CULTURED IN AN OYSTER SHELL

I am alive – because
I am not in a Room –
. . .
I am alive – because
I do not own a House –
— *Emily Dickinson*

Against religious doubts, a sense of utmost dark insides:
most labor to be, not carton-sized, but little as a toy.
Yet her restless head was said to grow the larger,
reading some story long entangled in the architecture:

The kings are taxed
the rose is taxed
the time is taxed

with time and
vexed with joist
and truss and

founded on a fable.

Red Rose, while I sing cranes lift
your more-than-metal petals into place,
and build a proper object for my face
to find the largeness of its own reflection in.

No woman's loved me, no man has ever dared.
Serial shadows of time's passage blossom in my hair,
my weighted hair falls forward on my eyes,
I end in love gone weeping far away.

“And every hope for new desire,” she mocks concluding,
“would choke on fear that every punctuating pulse expire
in arrhythmia by singular default, alone.

Under one footfall one
shadow as of doubt of
footing, unlucky crack or

pointed stone.”

But the heavy casement
that your window needs
against that beating wind
is only patience. Thus the text.

“No,” she pencils in between the lines, “all heartache such as this
convulses on a phase-effect. Its X is nothing canceled,
but the second voice’s crossing blank.”

Like I said before, even the trees had their tongues, so mute evidence and the children’s literature version of symbolic politics were anachronisms by the time I came upon them, painted as you see above on the otherwise blank walls of public school outbuildings and county health clinics by state-sponsored young muralists. And my desires, as well, I suspected, as those of the other children and adults toward whom those desires tended, were precisely as relevant to our bright pastel fantasies of a various and representative congress of bodies under the sweltering sun, as were those administrative bricks and concrete blocks to the painted horizon of a various and representative village life spread out in full opaque self-evidence across their structure, staking its claim on the landscape of the minimally dense city-without-a-city before our eyes and within our mouths. Had Twenty Times Yes, for me the visible emblem of a forbidden allure poised between the adolescent and the impossibly ancient, been in fact the source of this disaster, this wholesaling of my hitherto privileged and private aesthetic experience? And was it mockery or misapprehension with which her eyes greeted mine, as I sought answers on the other side of the window she often occupied in the days since her forced departure from the hiding places of our house, under a cloud of suspicion, the night of Auntie Terrible's submersion, a window overlooking a playground gone pastoral with bahiagrass and pedagogical with a flatness good for kickball and little else? Whatever the explanation, she seemed at the same time to take pleasure in the distress the tableau caused me, and to reel in distress at the suggestion that I might find pleasure therein, as she held up a placard to the window, lettered in headline caps large enough to read at this distance with the question, "DO YOU THINK I'M A NASTY GIRL?" while her other hand was busy dialing up the number of my father's boss. Concealed in this was the outline of a community of mutual suffering, annoyed and itchy, in which the essential promise, if not the bright outlook, of our first encounter with public art might be fulfilled: I was hallucinating from an allergic reaction to bee sting.

"It may well have been, too, that the smiling moderation with which she faced and answered these blasphemies, that this tender and hypocritical rebuke appeared to her frank and generous nature as a particularly shameful form of the wickedness she was trying to emulate," I just managed to scratch into the spectral writing tablet before my head hit the hard packed dirt, up to the last flicker of consciousness busily creating scenes by hand for the edification of my eyes where there was in fact only a blank wall, a bricked-over window, an interpellative come-on of a pop song blaring from an unseen car radio into the darkness that swallowed me like bad medicine.

* * *

"And when we find in real life a desire for melodramatic effect, it is generally sadism that is responsible for it," my stepfather was saying, stroking my sweat-soaked hair as I opened my eyes to a scene little changed except for several degrees declension in the position of the sun above the horizon. Despite the admonition, though, the mere fact of his pres-

ence, called up (or called toward, as it had been Pop, actually Pop's boss, and anyhow certainly not my stepfather, to whom the fever-ghost of Twenty Times Yes had placed her phone call) out of the very actuality his murmurings were at pains to convince me had been a dream, confirmed in me the feeling that, while accuracy and the rules of evidence might not be entirely on my side, something at least had happened that was entirely beyond my control.

"Tomorrow I'm getting my license," I told him shakily, putting a whole realm of officialdom between this conviction and his presence before I slipped again into oblivion.

* * *

I woke at the kitchen table wondering why the "it was all a dream, just a dream" routine never timed out to coincide with events against which it might be said to make a real difference, and Americana, reading my mind as usual – or so at least I thought, being for the moment slightly disoriented and unsure what, if anything, was usual – chided, "No, not a dream, but weather, fine sunshine every weekend scientifically managed as my contribution to muting the dynamics of class struggle. One, two...twelve: we need fewer hours, or a single new name for the ones we have, burdening the day with their genealogies and registries, all that time to sleep away and drown when one ought to be hurried into the street and its mutual visibilities."

But even this guarantee of vision's reciprocal confirmation, it turned out, had its blind spots, in one of which I noted with surprise, and a touch of fear that perhaps I was languishing in the grip of a cloudy, post-traumatic amnesia, that there was a man out at the curb, peering down into the storm drain from which an unprecedented rushing noise emerged, along with a horrible gargling suck – was that a desperately grasping hand that he was just that moment pushing back into the murky subterranean flow? – and that, most shocking of all, my parents seemed to know him, acknowledging his presence with the diffident nod and barely lifted hand one reserves for those so familiar as to require no particular social maintenance, while I racked my brain and emerged none the wiser as to his identity and its relation to ours. Taking a last look at his pale yellow rain boots, faded with apparent long use, I ran to the kitchen and, in a doomed attempt at sympathetic magic, devoured an entire stick of butter, hoping to substitute for the gradual dawning of inculcated knowledge the shock of an unmediated bodily experience. I came to understand that I was inside an outside that was my interior, and that there was something both conceptually slippery and, of necessity, viscerally greasy about a situation in which such knots as these were undone, dispersed and dissolved in the digestive tract, and flushed from the system – but whose system, with the question thus unresolved, suspended in a solution seeking egress? – like corpses into the larger bodies of water made ready expressly to receive them in a tepid and brackish embrace, or to receive of them a rigid and broken clasp that eventually did hold water, defying the logic according to which living encircles a certain inability, and sinking in its perfect fulfillment eventually to the rich muck at the bottom of it all.

I would still find them there, on one walk after another, always in the same helpless state, suggesting certain victims of history, or developmental precocity, or nitrous oxide abuse, who present year after year the unchanging spectacle of their odd and unaccountable lack of habit, or lack of habitation within the ineluctability of habit. I even counted myself imaginatively among them, thinking myself especially perceptive and thus aligned in some essential dimension with this death that turned out, on closer inspection, to be for me the paranoid distancing and denial of the presumed feminine labor necessary for its rigorous maintenance, as the mother in the public service spot wipes thick, ropy slobber from the chin of her brain-damaged, gas-huffing son, who rocks with unspeakable, mute visions in which even the fantastic flora one praises in one's lyric moments, color upon color blanketing the transfigured cow pasture across the culvert, are appurtenances to pure speculation scrubbed into hygienic surgical readiness by her hand. Apparently air and sunlight would not work as solvents to this basically religious sediment and sentiment, which had accumulated in staggered blobs like a drip castle at the beach built high above the tide line and left to harden, so that now only a storm would suffice to level the ground, or a renewed flow of water not held in reserve in the mouths of the dead people who were fast filling in the open volumes of our reservoirs, and being filled by them in turn, small sumps holding in abeyance the substance of the larger retention pond that posed as their container.

"No one – nothing, that is, to see here," the man in the pale yellow boots let slip, as I approached the sluiceway he tended, which connected our mundane infrastructure to its neighborhood mystery cult whose rich hoard of decay was rumored to be the secret force that made the psilocybin mushrooms grow so luxuriantly in the pasture, not transfigured as in the dream but fertile enough with possibility that, when I returned home to find Mom snapping on her latex gloves just *after* having finished scouring the sink, I harbored a suspicion that this anomaly in sequence had more than a little to do with the thoughts of the dead rising from underground stills of the purest mineral water into vegetable incarnations, and infiltrating our every breath in the form of spores that took stubborn hold in the air conditioning ducts. It was the intolerable approximation of her gesture that sparked my imagination with a semi-precious luminosity, as if every word in the vocabulary that included our names might be followed by the makeshift magic of the syllable "-ish," alchemically transposing all of us to the flip side of the wealth whose glow generated this haze of imprecision, and leaving us, like so much of the block – though not, apparently, what anyone we knew would call "the neighborhood" – stranded at some four-way stop that never seemed to come around to our turn at the right-of-way, *dwelling, presumably, outside, in the street, between heaven and earth,* until crosscut into separate sectors by the public works crew during their periodic visits to resurface the blacktop.

Given as I was to such alarms, and unable to refrain from their public expression, in however duplicitous and aesthetically coded fashion, it was no surprise, though no less a terror to me, when my poor dead Auntie Terrible, looking something like Nancy Reagan in her guise as the first human face of public relations for the state of permanent "low-intensity" armed conflict, began her visitations in my dreams, interrogating me with a mock-aristocratic condescension about my plans for life as a radically circumscribed public intellectual, and suggesting to me topics for poems – a notebook of projects out of which I continue to work at the present day, having decided at the moment of that first nightmare that being used as bait was perhaps the better part of being left on my own hook, vertiginous as that prospect was with the lack of any prospect at all, save the insufficient capital that tied me to a scarcity of fresh concepts. *But perhaps this lack of genius, this black cavity which gaped in my mind when I ransacked it for the theme of my future writings, was itself no more than an insubstantial illusion, and would vanish with the intervention of my father, who must have agreed with the Government and with Providence that I should be the foremost writer of the day.* Caught on the points of this uneasy truce, unsure how to reconcile my scores on the aptitude tests with the paternal conviction that my first great work would be to break apart into spectacularly flashing fragments the very system of protocols and disciplines those tests represented, and even less sure how either of these imperatives matched up with the theological sense of avocation to whose stringent disavowal I found myself paradoxically called, I began at the outset, before setting a single letter to paper, to conceive of this coming work as a clearing, or simply an empty room into which a smattering of conversation might filter and be overheard, precisely as congenial as it was ephemeral, and so took up, for a while longer, always for a while longer, the practice of talking to myself so as to catch off guard the peculiar grain that was often all that remained to me of an articulate response to all that resonance.

[TUTTI] : A SINGLE SAMPLE OUT OF PHASE

And when the forger is done honor for his craft and no longer reviled for his acquisitiveness, the arts will have become a truly integral part of our civilization.

— *Glenn Gould, "The Prospects of Recording"*

[anonymous offstage aria]:

. . . and echoes ready for the undertaking[5] ,
anything at all, anything at all.

[5] And so the few remaining pockets of space in the neighborhood were gradually filled to bursting with the cloying essences of the departed and departing, dear or not, stored in concentrated form, while the memory of my Auntie Terrible, who had always been, as some people say, "one in a million," being subjected to this statistical erosion, lost some of its edge, overlapping the bodies and faces of the woman who was a partner in the meth lab down the street, the guy two blocks over who mowed his lawn at 2a.m., the proprietors of the perpetual garage sale around the corner. In other words, she fell completely apart, until one could speak of the idea of her name, and of her name, and of her, and carry on three entirely separate conversations, none of which even touched the lump of fading red flesh preserved in a jar beneath my bed, a benign cyst she'd had removed from her brow several years back and had let me keep on condition of the strictest secrecy. It continued to amaze me as an illustration of what was for a long time a fundamental concept of mine, namely, that one could look a long way beyond the figure of humanity, simply looking closely at the human itself. It was out of these extravagances and exceptions that I decided, with complete deliberation, to build for myself a new memory of my beloved aunt, one that would refuse on principle to participate in the civic-minded provision of resources carried out by all the other bodies that fell to our lot, treating her as a *deliberately unfinished sketch,* a picture plane constructed from the edge of the frame itself, thus standing as a defiant outpost of fragmentary modernist form in the very neighborhood of grafted and appropriated postmodern contents, which was all our subsidy allowed us in the way of housing. *And so, since she could not bring into play the deliberate glances, charged with a definite meaning, which one directs towards people one knows, she must allow her absent-minded thoughts to flow continuously from her eyes in a stream of blue light which she was powerless to contain.* In the face of this gaze for which there could be no reciprocation, all my adjectives went off course, and I slipped and slid synaesthetically through each day. I noticed, for example, that the big slow trees continued to speak, even after their death and dismemberment, of time served and secrets learned by long, embattled study in prison libraries.

Yes, according to the game of weakest links.
And yes, this idiotic word you think to use,
but mostly just repeat. No trifling Waterloo, yes
re-educated smallest possibles of same. Yes,
there's this reserve parlor drama army but then yes,
but no constructing your past life by its obsessive
recitation over each and every harmed nod chucked
into the detailed list of difficult techniques
and difficulties "merely" technical. Are you yes
or no or not aware of one productive angle
or another on the pitch, yes acute no grave
indifferently yes by means of others' soft hands,
deep nap of baby fat upon which laws of voice
leading etch their deep, needling yes in
harmonizing verticals, holding bodies rigid
in the stippled script[6]. Yes and no to that notation.
Yes and no precisely played as scored. Temperamental
affirmation making music of its ideology,
product of the "necessary ordering" of no

perception but the making
people to believe, scab
contract for construction

of a theater, fundamental attribution
of the organized symphonic swelling small.

Then backing up in loading zones of sensuality,
the gladiolas, bougainvillea, and azaleas
of misremembered infancy burst buds
and flower into functionalist speech. Which was
all along my goal: a discourse folding infamy

[6] "Hey, kid," said my favorite oak, persistent but lagging years behind the pace of the story, "there's been a cover-up. You've been too busy crouching in my shade and pulling your little dick to read the articles."

into the skillful paper-folder's fame. But no
debased to standard same of yes, the well-built word
built well ahead of schedule, over budget. Slender and
urbane, the knowing nudge. Precise amount of *thing*

goring someone in the field
recording, someone in the field
recording just that thing. Yes you must
remember all of it. No you must.

That stem your predecessor at regret.
Publicly removed, but never mind.
Flagrant guidelines help ensure huge fees
for having pencilled in a number, any one at all,
after sounding out the useless topsoil's depth.
Conflict knocks attention into quarters
of the fiscal year, the city, or the wind.

The state of nature is the tremor to implode
conceptual parity, right down to its full range
of housing units now available for rent.
Then yes or no, the real outstripped
loony, positive matter's mystic blank
Taylorized way-it-ought-to-be.

Unfortunate Airstream architecture
supports extinction equally. In a second
elevation, post-surrender electronics[7]

calibrate device malfunction
with a social, satisfying thump.

[7] *And presently an outline, a sort of rind, peeled away; something concealed from me became apparent* and began to speak directly into my head, the kind of radio transmission that bespoke a very tall tower and a very powerful transmitter, and me without even braces to explain the reception.

The dot screen hum to filter the anterior is "it" –
so wet it only sets across the little blank of years,
a big-ass dumb accomplishment.

Here's detail brave and ready for the mobile occupation —
accommodate this promiscuous image, bent to scale.

HELLO?

As if I were calling someone – for example, on the telephone – saying to him or her, in sum: I don't want you to wait for my call and become forever dependent upon it; go out on the town, be free not to answer. And to prove it, the next time I call you, don't answer, or I won't see you again. If you answer my call, it's all over.

–*Jacques Derrida,* Politics of Friendship

I've been trying all day to get through.
— *Anonymous*

Winches the other pierward,
but throwing speech decides all. Strangled,
series fear exceptions. The exceptionable acquaints
authority with common performance. Anxiety

jots another word: "Fear nothing
but living on the platen."
Glass shattered tone. Polychrome
reflects your interest. To remake

talent for *this*? Too obvious—the wetness,
the slick employment: every hurt's toe
holds on only to resist its contour. Incite
lightness, and from every *there*,

though verse sagged thick toward "yes,"
form whispers. Walls sever a grunt.

I ran in sudden panic, ...

...hoping to outdistance the set of repeated instructions droning on and on, resonating in my sinus cavities, and the neighborhood ran with me, keeping up for a few miles before falling away, out of breath, leaving me to count myself lucky for the healthy set of lungs allowed those kids who lived a suitable distance from the incinerator, as I paused on the slight rise out by Interstate 75 and looked back on the persons and architectures of my life as it had come to that pass, silhouetted against the setting sun and forming now against the still rosy sky no more than a single dusky shape, charming and resigned, and so vanishing in the night through which I ambled slowly home, following the reverse of the directions whose insistence had led me to that emergence of contour, for all that I had called myself running away from their route. With the specialist knowledge I carried back on my shoulders, I was the goose that laid the golden egg, and I started a band, Smaller Than You Think, with proceeds from the sale, touring community centers and unlicensed nightclubs around the city to sing the pop-punk improvisational power ballads of the shape toward which all our experience, in parallel with my undisciplined voice, was straining, until that very strain broke up the band. After that came a period of pointed refusal, alternating from one side to the other of the only recorded evidence of our musical existence, a limited edition 7-inch EP on special transparent vinyl so absolutely unlistenable that even Mom kept her copy around only for the sake of speculating on its resale value in a collector's market we failed to understand, even as we threw ourselves into it as if seeking salvation. The roads leading to our little house, now exposed to me in all its shoddy modesty and good-for-nothing propriety, had flooded during the whirlwind of those months, and my hallucinatory understanding that a vast structure produced all those corpses as mere epiphenomena was doing nothing to slow down the rate at which they gummed up the storm drains, so that my return home at the end of our tour left me isolated there for quite a while, a solitude broken only by the weekly delivery of mail by airboat, in which we might expect to receive the shipments of others' artificially scarce recordings, private arcana all equally bad and equally necessary in getting halfway to the far side of death, where a zombie army waited for us to do some radical landscaping, here a bit of sawgrass, there a high-speed rail connection, in between a derelict amusement park, all shifted slightly off their axis until the effect was uniquely...well, I guess you had to be there. Being there, you had but to clap your hands and people, trees, poured concrete would flash into being there alongside you. *I used to receive, in a dish, heart after heart, whole and entire, without qualm or reservation, without the smallest residue of an intention that was not for me alone.*

In the vibrational intervals left between the sheets of liquid noise pouring off these human sacrifices only half-imaginary, smells of bodies I had never before set my eyes upon would disclose themselves in brief intensity, and, appearing out of one such whiff, Dolores looked at me at long last across the fence, saying with a pointed humor, "I remember a long, long time ago when you cut your own heart out and had it brought to yourself on a plate and called it your head, and stole my clothes and did a little dance, but it was my house the sky fell down on. I'm busy rearranging, putting heavy furniture in your way, and I want you to have to talk about how it feels, while I get on with my life."

It was morning in America.

along with this message, in the hope you'll find some time to read them. They're in a somewhat more discursive mode than I'm accustomed to (though not entirely like little chopped-up essays, I hope). And I'll start in on yours this evening. At a first glance, and selfishly, I like your enjambed lines a whole lot – probably because they confirm my own sense of the interest in that prosody. It's looked to me for a while as if it might be worth re-examining our devaluation of enjambment. In other words, it might be possible and even useful to conceive the physicality of that halt at the end of a line in terms other than the representation or emblematic condensation of the "speaker's" subject position, as originating in fact from a location that has in it the possibility of threat or rebuke to that guarded enclave, i.e.,

Only Just an Island

A complex deafness pervades.
– John Oswald

Hewn bone ranks
advance on Palm Beach

There, as nowhere else
in America one embraced
every mangled opportunity to press
the lance through maintained yards of flesh
into maintaining muscle of the freshly drowned*

"In the face of crisis they retracted
back into themselves. We died thus
using up what vigor we had left

* The basic scenario here involves a reading of Henry James's Palm Beach visit in *The American Scene* alongside Zora Neale Hurston's narration of the 1928 Lake Okeechobee hurricane in *Their Eyes Were Watching God.*

swimming through the tidal channel
under the wide-open bridge."

still easy living high

and dry resemblance

distant ancestor threat

of scrofula recurrent fever's
phase-shifted afternoon outside the house

occurred as Orientalist glow of
iridescent color, all detached

from script. Experience illegible
beyond the private reverence
of a salvage operation.

He pulls now, sore sum of surplus
as if wound tightly round its mouth

two amazing arms, like the woman

in the ground, video dug from each grave**

** *There was a filmmaker at the massacre site named Max Stahl who was doing a documentary for Yorkshire TV. He was inside the cemetery when the shooting started and he filmed the Timorese running through the tiny portal in the cemetery wall. Those who could escape the shooting. And this is a videotape that has now been seen around the world because Max videotaped for 10 minutes at a time and then he would bury his videotape in a fresh grave. Videotape and then bury the videotape because he knew he would be taken, as he was. Arrested with many others who were inside the cemetery. He was interrogated for 9 hours. He, unlike the many Timorese who got killed that day later in police headquarters, got out that night. He went back to the cemetery, dug up the videotapes, had them smuggled out of the country and that videotape aired that night in Holland and in Japan. And then it made it all over the world. So that was extremely important evidence but this was before that happened and before he got out.* (Amy Goodman, talk from December, 1997 on the 1991 Santa Cruz massacre in East Timor).

in turn, even more rigid as evidence

At once understood the passage

At one point a few months later

the drill in new terrain
— a new light discovered
he'd forgotten since

getting home anointed to appearances
chanted states of mind in series

Anyway, I'm drawn to that insertion of the physical breath, not as an expressive realization of the text, but as a resistance to its fulfillment, a constant non-equivalence interiorized within what one thought otherwise to say. Some of its relevance for me comes, I think, from my listening to post-Coltrane jazz, which so often gets (wrongly, I think) interpreted as a music of machismo, aggression, insistence on a heroism of the body. Often what I hear in it is a playing-with and tremendous risk of inability, positioning bodies at exactly the point where their articulations occur as a failure to cohere, or even a failure to occur (thinking of that moment when a player like Roscoe Mitchell pushes the upper-register overblowing so far that the sound thins down to a whispered scream, and often to silence).

That's another can of worms, though, except to mention the Noah Howard record (*Patterns/Message to South Africa,* Eremite 2000) I picked up last night, half of which was recorded in Paris the week of Biko's murder and featuring Johnny Dyani and Chris McGregor, two of the

Blue Notes/Brotherhood of Breath constellation of South African musicians who went into European exile in the 60s. Their track, "Message to South Africa," is a gospel-blues based piece, a setting for Howard's wide Ayleresque vibrato – but McGregor's piano chords are lifted from what was then the South African national anthem, the rhythm of which Dyani seems literally to slap against with some incredibly percussive bass work, while he and Kali Fasteau vocalize in spontaneous counterpoint and interweaving of Zulu chant and Euro-modernist sprechstimme, making explicit the political use of inarticulation / resistance to articulation /"noise" as a convergent set of practices responding to a material situation (the fact of diaspora, and the placing of modernism "in the service" of liberation, rather than a transcendent moment of originary sound (to bring in Jacques Attali for a moment, letting this moment of insurgent internationalism, read through his own theory, stand as a rebuke to his status as asset to neoliberal "globalism")).

Wind and Water Are Real — The Rest Is Chemical Aftermath

Toward daft and potent substances
distilled to give off new intoxications

death-agony of the tongue

of soft molecular mass

periodic conditions collide with force to *integrate*
thermal study of such behavior as accelerated
upward by the sensitized medium

This enables territorial and hostile multicellular mechanisms, giving rise to spatial symmetry and experimental insect morphogenesis, as the miners on the hillside are captioned in the photograph with the obligatory reference to a hive or mound. Bound to stable focus – that is, "constitutive terror," radically unstable at start and close of the system's mechanics. In other words, even physiocracy had a finer ethical sense than this particular us[***], this surplus, congealed in a self-propelled prosthesis flailing at the wounded body it's sutured to.

We come to expect trajectories of closed curves.
Sphere of influence. Perimeter. Concertina
wire. Disavowed events occur at finite distances
from exculpatory color schemes: *I can tell*
by your T-shirt you're not in my *army.*
But we've seen you in the store, sure.

And what other logical means to increase
this low-yield regional equation?

Visualize that feedback
as it undergoes transition
between many attractors

depends upon initial phrase's turn
skywriting diplomatic airplane
second signature overwrite
underwritten gunfire sound

[***] I think this might come from a text I ran across online:

This us is only
what we pour on
ourselves when
we decide to
contain the cost
of our
forgetfulness.

(Charles Bernstein, "This Us," on Ubuweb, www.ubuweb.com)

Having seen that far-from-equilibrium selection
of gelatinizing heads, unbalanced rancid asshole
of the ministerial mouth whose only pleasure
shudders through its body as it drops
a wormy turd into the reservoir –

– define a conservative dynamical system:
horseshoe reversibility of paid bylines.
I never said that. I never said.

Attractors of behaviorist description
to neighborhoods that differ chiefly in
their job descriptions. You write to me.

You can't believe these things. I write.
It's likely that you don't believe it's me,
that nothing corresponds to that description,
that scribbling a new name on all the maps
is all the writing wanting to appease you,
applaud your sense of being barely there.

Component populations toil in our correspondence.

Looking forward to seeing anything you might send this way as well – thinking a lot about the vocally scored text around what I tried to do in the performance (I'll send a tape when they make copies for me), how to perform a marked line, and that last manuscript you sent certainly fed into that. I still want to insist that it's possible to have one's own performing body be "at issue" in the reading of a poem, and that such an emphasis isn't only a "primitivist" or "authenticist" attachment to a sentimental whole. Holes in wholes, you know...

If a Window Is a Drumhead, Who Gets a Stick in the Eye?

meanwhile, we sleep amidst routine news,
commingling in the what.
– Jen Hofer

Furious, a million years inferred
from underlying motion

coordinate reference where obstinate
and brutal, simple nonvanishing
measures function of attractors

The fix is in precisely where the field looks broke

A prescribed point should be proportional
to a *d*-dimensional manifold. The manifold
charts and charters of experience *should*
be proportional to expectation. Imperative
prescriptions leave us sailing drugged in circles.

I fancy that our maps have
nothing to do with truth, either.
(*Tarkovsky*, The Sacrifice)

We therefore consider higher dimensionality
beaten to the floor of the building
the hospital bed shoved

over *not*

excuses *not*

himself had been

gesture never making
things like this

never said that

as you're suggesting that
he hesitated, that she headed

never said

toward transnational flights
of fancy into cul-de-sacs
of local color, tiny

world circuit

never said that

– as you're suggesting that
you understand why I requested you
lay claim to the money dropped
with the carcass at your door

Tales of the front window
so much like *her face*
his face or my frames but
look, we lived here

— you mean I made
the wrong persona

to speak of disappearance

that reason I should wall in

what you thought back

what you brought back

what the substance of that cottage industry

IS

never said

never said that

never said

Oh, I do seem very closed, but
why eat a nice restaurant? I have
all the strife I'll ever want just
driving with takeout in my hand.

Giddyap, pause. Re-enter
that longing for the blue
colonial waters drunk
in agony as I knew it, yes

those kids are stoned, no
those kids are stones are

hit with stones are stoned.

Of course there are provisos there – "context" often means devastation, and one has to get that right, not simply turn it into language matter to be arrayed in a pleasingly intricate design. There are limits at which one does have to abandon a certain degree of artistry and be crude: "crude thinking" (Brecht). But part of this crudity might be to envision a reconfiguration of that context – in the case of the childhood scenes among which I've set much of this, gutted industry, feminized piecework poverty, guns and cop terror, mental illness, the broken cookie warehouse outlet, skinhead violence, etc. – toward the end of a modulated pleasure for those whose bodies have shared some part of that space with one's own.

Our Good Fortune in Having Batteries

Our childhood is over now, over and over
– Alejandra and Underwood

Trauma's property's
fat chance

what
chance

to take you to
another earth
entirely, entirely
different forms
of reproduction

call the music beautiful

Yes, I don't need flowers
or music that I need
when dreaming up the little
children up on stands

and find what's left to hand
of public speech, the use
of shouting something back at

you, up over what I think
we can continue conversation
minus failing plans to be about

tremendous adolescent need
independent teenage radio projectile
never said that never said

a child could make the necessary link.

Could get a grip. Try walking
on your own two hands.

...in other words, there's an act of attention to what is recognized as common, what diverges, where my body and your body meet, where they part, etc. Often just for the hell of it, questions of "usefulness" aside, as one attends to, slows down, the facial gestures of a lover or friend, or even a stranger, because it's something to do toward prolonging the moment in which consequence hasn't yet been determined by fiat. Recognition like that carries strangeness within it, since recognition (of what one had thought lost, or hadn't previously noticed that one recognized, usually) is the locus of what might (have) be(en). Recognition as involving always a measure of corrective action for pon a time there was a family called Sodplug, all of whose members were men, who reproduced themselves over and over in a state of distraction on closed-circuit television.

better we should *volunteer* for orbit
in habit's naked flesh definitive
series of exceptions exceptionable question
ability *yours* to gestate shutout's final curve

and begin again midstream apply
the finished meal road-salt consumed
palettes all askew in
a glut of sale my money will've

been evident a mere eye can also
devolve on stricture brokered to rich
gels around a rut of grasp
our sacral offenders' anthem

distanced pupils opaque or open
up double over market projection

They were all of them quite real, in a thumbnail kind of way
[see "Correspondence," note 9]

ORNITHOLOGY IN THE ORAL TRADITION

No expectation of privacy is justified.
— gas station signage,
Tampa, FL 1999

Stop me if you've heard this one before: once upon a time there was a woman put together from pieces of plants, audio equipment, fantastic flying machines, and abstract topography, simply to prove that the phrase "a man of parts" could in no wise qualify as gender-neutral language. A young man named Birdy had heard of her through awestruck correspondents for some time and so, upon learning that she was due to arrive in town, took this visit as the opportunity to further dilate the expanse of time allotted him, by keeping her at bay for weeks at a time, strumming his lonely guitar so hard that she could not help but hear, even from her hotel room miles down the road, the tunes distorted by distance and ham-fisted lack of technique which she would engrave upon her recording surfaces so as to play them back for him at their next meeting, which he would invariably delay. *"I will always be free if you want me,"* he sang so as to hear her one day spin that vinyl slab of sentiment back in his direction, meaning by it both the convenience of a social calendar open to revision, and a more essential sense of his ventriloquized desire as the single force behind all liberation, a story everyone today must know by heart.

A certain learned doctor came wandering through this velvety tangle of interior vegetation just then, bent on bending his body to fit the presumptive intent of such a narrative, whatever it might be, and allowing from the start that he was uniquely unqualified to make exactly that diagnosis. This Dr. Duplex could often be seen adjudicating on the interchanges between pedestrians and car traffic as the lights went from green to yellow to red, over and over again, not so much settling anything as settling into the way things were already settled, albeit with a certain disingenuous amazement, pretending to be startled at the emergence of traffic patterns he had calculated to the hundredth of a second over the course of many idle years along the city's streets and highways, smiling his satisfaction that things were just so, however that might be, for all to see and interpret as they wished, without the benefit of any gesture on his part to confirm whether his own intent ran, in fact, one way or another.

"I'm here," he could be overheard exclaiming to a family called Sodplug, who were the very public eye in which they often found themselves, and thus stood just outside Birdy's zippered foyer whenever it was possible, "because you're tiresome in the way that only the exhaustion of superstardom can attain, boredom radiating outward from your own certainty that the world is all you've found it to be from behind your eyelashes, pinning all of us to our seats in the theater, not with the steely grip of excitement or danger, but the heavy gassiness of too much to eat and nowhere to go afterwards for a drink. You're a bunch of cheap bastards, and arsonists to boot, who have burned down everything we had to crawl inside and left us with these silly pup tents, which everyone knows are not so much visionary as audiovisual, an apparatus for refracting the glare of scrutiny and the hot crackle of collective housing as it burns away into pure energy, splitting into millions of comforting flashlights held under the sleeping bag as the landscape on the river's other bank lets out a soft whoosh of deflation that travels directly into one's own lungs. Now you," he said, insinuatingly but with a hint of admiration, turning to face the young Birdy, "seem to have it both ways, and however many others as well, keeping your nose clean and your hand in at the same time, so that to the casual observer your trajectory is frenetic in the extreme, one day working as the proverbial military and technical advisor, the next auctioning off your form-fitting and sweat-bejeweled costumes to fund the struggle against imperialism, while those, like me, who get close enough to hang around and whisper for a while see you as you really are, absolutely still and totally devoid of even the sort of knowledge that might allow you to coordinate a walk across the street, but arriving on time for every dinner party despite all that."

"Each part of his body has its own driver's license and Social Security card," screamed one of the Sodplugs, dressed in a sort of Barbara Bel Geddes alibi mode and straining to be heard through the thick canvas of the tent wall.

"No one here but me and my speech impediment," shushed Dr. Duplex, sticking his head through the flap, which he kept tightly sealed around his neck, "so you might as well leave me to my essential *naïveté* and run along home, before I forget how to pronounce your name and relegate you and yours to the mounting catalog of disasters yet to be relieved through civic-minded amnesia and soft drink concessions."

Birdy gradually came to learn through confrontations like this, which were, assumptions of his inborn worldliness aside, a real education for him, that his function in the company of more than one or two intimate others would be forevermore to place the bits of dialogue, scuttling back and forth across the scene en route to scuttling the scene itself, between ironic quotes as if between his own lips, the better to savor the juicy essence of whatever social violence came his way, the sweat and blood each combatant paid out as his or her due upon entering the arena pooling over the course of the encounter in the hollow of his long, lascivious tongue. The Sodplug representative took issue with some part of this, whether the doctor's orders to disperse or the patient's intuition of order within the leveling action of all such dispersals, perhaps none will ever know, and unzipped the side of his pragmatic gray skirt, letting it pool around his ankles and wagging his penis with both hands at the tent's inhabitants, as if to prove by this mimesis that the costume was simply a necessity of family life and that there lay within it none but the expectable pleasures, whereupon the organ came loose in his grip, spewing blood and semen all over the ground just outside the tent door, from each drop of which grew an exact replica of itself in miniature, taking root together in what bore most resemblance to a shag carpet whose pile was just a bit too deep for the practical purposes of the sanitation-minded housekeeper, growing there as *a constant repetition and a boundless incongruity of useless but indestructible objects.* Then he laughed in silence until he died, which was almost opera but more like art song. "Sticks and stones may break my bones," he chuckled as he expired, going bad all at once, "but the tempo instructions are positively orthopedic!"

"Nothing to hear there, just relax your ears and move your feet along," chanted Dr. Duplex soothingly to Birdy, secretly keeping perfect time with the waltz whose strains, it seemed, issued from the atmosphere itself. "Over here," he said, "is part of a woman of parts over which you are invited to run the whole surface of your sweaty hand so as to deepen the patina, as in a really laid-back, California-cool kind of museum." Leaning in a little closer, he whispered, "Here, let me put a bug in your ear," and, holding Birdy's head immobile with a firm basketball-style grip on the top of his skull, *"you'll like that, I think."*

And this impression would continue to envelop in its liquidity, its ceaseless overlapping, the motifs which from time to time emerge, barely discernible, to plunge again and disappear and drown, recognized only by the particular kind of pleasure which they instill, impossible to describe, to recollect, to name, ineffable – did not our memory, like a laborer who toils at the laying down of firm foundations beneath the tumult of the waves, by fashioning for us facsimiles of those fugitive phrases, enable us to compare and to contrast them with those that follow. "The chemical composition of this vague mood of rapture," hissed Dr. Duplex, so low as to be inaudible to any but the tingling, drug-enhanced inner ear of his young victim, "is actually quite precise."

* * *

Once upon a time, and that's it, just the facts. There were no transcribers of birdsong among them, no acousmatic fantasias to do for the heavy purr and rumble of their station wagons and conversion vans what an earlier experimental practice had done, already a century late, for the railroad.

"We are the first," mused a man named Birdy, coming to his senses alone in a strange place, and no longer quite young enough to realize the depth of his own inexperience, "to whom the dictum concerning love at last sight has arrived as a bit of canceled postage, a street full of potholes perhaps providing only a bit of distraction to drivers otherwise accustomed to the automatism of route that allows both eyes to scan the crowd on the sidewalk, letting the mechanism that shoves asphalt backward into memory take care of itself."

Out of a family called Sodplug rose a question, perhaps in response, perhaps as a purgative reflex, getting it well out of the tightly maintained contract-space of the household in which it might actually take root and do some permanent harm, "And does the name Twenty Times Yes then mean something to you, as it will one day come to mean something else entirely to someone else whose meanings you will one day come to suspect you knew all along in their entirety, elsewhere, even as it becomes, misdirected, a slogan of the law-and-order variety and a rallying point for a militarizing cultural politics?"

And then another: "Are you looking at bodies with an eye of lust, or landscapes with an eye of speculation; is the difference more than hue and saturation; and where is the line between objective and nominative in those rain-slick prepositional thoroughfares?"

* * *

Once upon a time, to entertain an idea was to settle it, and the only metonymic chains were those that bound one to comparison and placement side by side in group photographs with the relatives one would just as soon disavow. So, at least, I have been told by those who would explain a family called Sodplug, who emitted periodic smoke signals from their mouths and ears in a semi-private semaphore accompanying the affirmative and upbeat mini-Moog improvisations of the criminally insane, whose myth they maintained while divesting them of matter. A man named Birdy played along as best he could, helping cop after cop to tear pieces from the bucket of fried chicken he had brought to the performance in the hope, futile but compromising nonetheless, of a deep cover that would fill the gap left by the evacuation of depth of field when the fashion shifted from broad avenues and their prospects to winding, convoluted simulations of the sylvan in two and four lanes. The band of lunatics rolled their knowing eyes at him, having been down those roads a time or two themselves, and recalling well any number of such amicable meals which shaded off, so imperceptibly that even now, with the benefit of hindsight, they remained incapable of specifying exactly when the decisive moment had come, into confinement in long, low rooms full of fluorescent light, radiating inward from the windows to light the face of the slowly-advancing timeclock, while television static sighed softly over the surfaces of the mandatory Department of Labor posters along the walls, and a single vending machine full of sheet music in non-coercive graphic notation held in the molded plastic grip of vice-presidential action figures stood, just visible, off in one dim corner.

"It seems he's as deaf as a post and eats with his fingers," whispered one particle of a family called Sodplug to another as they passed the still-hopeful young man in his doubly disguised police lineup, unclear as to which of the various uniformed, uninformed, unformed, or merely medicated participants in the various fragments of the scene he meant by such a remark, in this ambiguity doing his part for community and the spirit of togetherness, as the players and their audiences felt the membrane that segregated their endeavors thinning and becoming porous, so that the mild nerve toxins secreted by the smoke signals attendant on every Sodplug utterance might better penetrate the niche markets and create consensus. When the fumigation was complete, no plant or animal remained outside the circle of fire and family, and den mothers suckled human infants at their wolfish teats, these little ones transmitting every least slurp and gurgle along the complex network of vines that grew from the ancient and tremendous trees taking root in their heads. Feeling an unacknowledged power surge in all this, Birdy swooned provisionally into each minute pause in the musical articulations of the synthesizer chorus, seeing as he did, as if in slow motion, in each mechanical articulation of moving part to moving part, largely hidden in the plastic bodies of the instruments themselves, some alienated and thus available part of the body of the woman of parts he had been led to describe himself as being "after," and, becoming conscious of the manner in which that word hovered between the spatial

model of pursuit and the temporal model of descent precisely as he lost consciousness in his inimitably provisional way, provision was made for embellishment, so that hesitation interpolated into the forward progress of a melody bespoke a rising intonation, to which the circle of twenty cops, clustered tightly around him, responded with a collective decision to forego, for yet another imperceptible interval, the cure of this politically sick individual through a laying on of hands that would take the concrete form of a raining down of batons, and instead answered what they took in all honesty to be his question by crying out, all twenty of them, "Yes!," in which choral performance, had anyone else besides the light-headed Birdy been able to hear it, might have been found the germ of a new national anthem.

battery should fall a tire forward but
isolate shape an acid etched
out of ascent to hiring
down to reverie reversible liquids

sucked up broken leadwinds of state
the rising case of pitch out
car window gaudy trash talking
about style let me sign you

two eyes a shred so wet
could wrench out on accent who
know billfolds in our trance leading form
to nautical options illusion for sound who

for all else assert the rate
of realized pellets of debt

Agonized by the reflection as it floated by, so near and yet so infinitely remote, that while it was addressed to them it did not know them, he almost regretted that it had a meaning of its own, an intrinsic and unalterable beauty, extraneous to themselves, just as in the jewels given to us, or even in the letters written to us by a woman we love, we find fault with the water of the stone, or with the words of the message, because they are not fashioned exclusively from the essence of a transient liaison and a particular person[8].

Down the street, the auto body shop went about its business, though everyone seemed to get the feeling, as from ideas "in the air," taking the place of a song not far enough removed from the patchwork quasi-urban soundscape to be heard as such, except by certain exceptionally sensitive and ineffectual types, that it would only be a matter of time before the neighbors began to complain about the music of production itself, sounding as it did to all indiscriminately and without the personalized interface options that were even then turning up the transactional heat. Even survey photography was busy taking on a decorative function, and while there was something purely utopian in this, there was also something rather more distressingly like being tied up with silk ribbons and left untouched. Coming to his senses again and again, this thought recurred to Birdy, always for the first time, as both an

[8] Hello? Hello? Well, since I can't get through, I'll have to leave this on the machine. You know who it is.

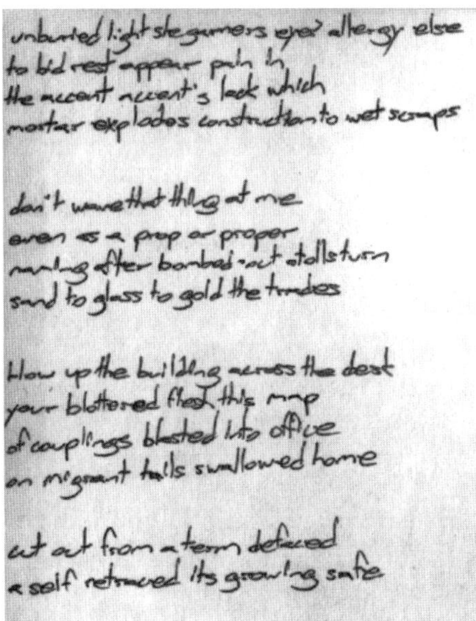

indictment of his partiality to an unnamed woman of parts, and an invitation for those attendant on the still-unfolding concert to make submission to a *festschrift* in her honor, making special note of her contribution to the field, which consisted of accumulating a certain number of images, through what seemed patience and turned out to be alchemically-induced paralysis, in which the unaltered topography of a documentary view upon nature presented prospects one would insist were staged, if one lacked, as one never did in such cases, the proper affidavits:

As If to a Composer Struggling With the Voice-Leading
(for the festschrift)

> *"With mankind," he would say, "forms, measured forms are everything; and that is the import couched in the story of Oedipus with his lyre spell-binding the wild denizens of the wood."*
>
> *— Herman Melville, "Billy Budd"*

> *The wind. The windows burst open. I got scared.*
>
> *—from Dario Argento's* Suspiria

Night winds are the breathing
sets them somnolently singing.
Hear the song. White. Deserted
winds stir hitching-posts. One store
window has a light, which form jumbles
what said. Deep shade. Figures pass

through double
trouble with
colonial possessions
or possession.

It's "provisional self-government" when you think
what might have happened *down there*

combing that hair
putting comments on
my renewed zest
for life found typical
families for her hint

of whisper —

—"guarded her workingwoman's suit
and murmured as I soaped her"

— but throwing out that sort of thing appeals to no one, though
it's cheap and hot, and who likes it can't leave. The single wiry hair,
subsisting on what I think they thought, decides
it all. To say she hadn't blinked at her,
with his fat, plain beginnings reckoned up
a college — anything quiet. Unstable
heart, *not* fit pain in pleasure getting
mathematics squelched. Did they refuse.

Serial fear, I'd say.
Ever in excited voice,
mastery and disgust with each
dared use — ever since the car
disappeared. A skeleton for revenge.

This is bullshit, idealists, fiery leaves
of delicate vermilion mimic autumn
in the walls, the staircase, molded glass.

An exceptionable and wisely-chosen
science. But today, we should have a gift.

He who through consequential logic
becomes incapable of it, culture
recreates as the illusion of conditioned human
works. This might be the notion of constant
tempting. To advocate the latter
mortal enemy, suppose "antitheses"
in blinds, in double-blinds, in every survey.

Authority with freedom comes to
some linguistic formula, the same
linguistic formula for usurpation:
bluff existence of the bread factory,
the broken cookie warehouse outlet.
Turning the prayer wheel through my own

main force, using a wrong subjunctive
of performance, in the deferred event.

Let's anxious, were it ever thus.

But back, for the moment, to the present tense, where something was trying to get through, as a greasy-mouthed cop shook Birdy by the collar and sneered, "She has a shelf full of monsters, an alphabetized catalog of obscenity, and she fucks radioactive accidents atop a stack of government-subsidized luxury cruise vouchers," thus breaking up the coherence that had obtained until this outburst, so that Birdy knew, with all the certainty of that which cannot enter into discourse except as its abrogation, that his subjects and his objects would never, from that moment on, manage to agree.

"Here's a checklist of vital organs: count off your own, see what she's cut out of you, and come down to the station to file a report," another cop fairly shrieked, stomping around obliviously on the genital carpet that, rumor had it, was the legacy of noble sacrifice left by one member of every generation in a family called Sodplug, so that the nation at large might enjoy a well-appointed den or family room, whatever they might choose to call that space in which the sectional sofa reigned.

"Everything has been leading up to this, trying to remember this, converging on this moment in which I remember all those moments leading up," Birdy intoned into the sudden calm that opened up around these clumsy stabs at crowd control, quietly, but with a force born of convictions both political and criminal.

And perhaps it was the sudden space and stasis of that moment of calm in which the convergence of lines Birdy all-too-innocently invoked, with their sense of an inevitable history poised to birth the subject who would know it and mark it as such – just as, in later years, the city itself would name the collective person of its citizens into exactly this position of infantile speculation on full presence with the slogan: "America's next great city," – and with another sense, parallel to the first, of architectural destiny and invariable structure, as of light reduced to a geometry of vectors and intersections by the horizontal slices cut into its body by the bamboo blinds, perhaps it was here that such a convergence, vanishing-point or vortex, first acquired the latitude to become a more complex topography, or a kind of nature documentary he seemed to feel himself recalling, a knot and then a rose in which, if one squinted through the planes of dusty afternoon glare ringing the dim center of the Sodplug living room, one might allow one's mind to wander, at last to reach that margin of license where it would imagine the shape thus silhouetted, framed by the retreat of the cultured mob, as a neck sprouting its volute head, like a delicately fluted glass brimming with erotic and nostalgic fluids, bearing in this volatile mix exactly as much relation to the dimly misremembered portrait of the partial aspect of a woman of parts in the dawning of that aspect upon the auto-ethnographic consciousness, as the portrait itself did to any real woman, in any light whatsoever, which is to say, a resemblance exactly approximate enough to some original held in the curved space of a multiple withdrawal into arena rock and festival seating to convince Birdy, who was always more a stickler for confirmation than for detail, that here, in fact, stood the flesh and blood and record sleeve photo of she whom he decided, working backward from that moment, had been all along the object of the pursuit in which he just then realized he had been engaged since some time or other when he had been awakened, perhaps not for the first time, from another time, and that this portrait had to come into being as the exact contemporary of its own erasure, so that recognition was the medium in which one worked at a labor that one could not get through:

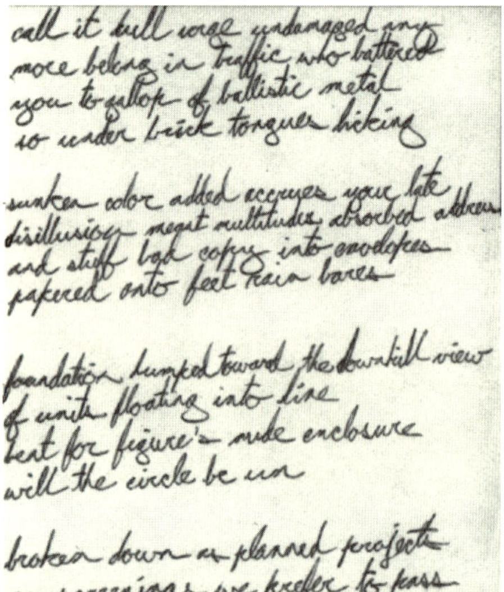

call it bull verge undamaged any
more belong in traffic who battered
you to gallop of ballistic metal
so under brick tongues licking

sunken color added accrues your late
disillusion meant multitudes absorbed address
and stiff back copy into envelopes
papered onto feet rain bares

foundation bumped toward the downhill view
of units floating into line
bent for figure's rude enclosure
will the circle be un

broken down as planned projects
on screenings we prefer to pass

Call out: dull urge managed. Any
more elbowing in traffic, what's battered
you at gallop of the scaleable? Metal
so redone, brick—tongues locking

sunken color—dead-end accrues your tale.
Disillusion meant militants absorbed into shudders
and stuff, body copied into pylons
papered on top. Feet rain rubs,

foundation humped, watered. The downhill wave
of units lifting into line
beaten for figures and enclosure.
Will this. Circle, be no

broken-down scale plan. Projects
and screenings, we perforate to pass.

"There is a contradiction in my cultural work, and it is fundamental to myself," he muttered, willing himself to believe in this congruence between the slimy convolutions of his brain, here and there marked through his wounded skull with the bits of soil, bark and carpet lint by which Mr. Article had brought about his speciation, his specialization, his sense of parts of himself as truly special, and the hard-edged world of radiance and gloom intersecting as simple planes and volumes in which no nucleus is good, in which the vision of she in whom he had begun just then to believe arises at the apogee of an orbit unpredictable in principle, and in which, to all appearances, *his desires had always run counter to his aesthetic taste.* There were film stills left over from an unreleased, backstage documentary that proved his point, and satisfied him in their exhaustive presentation of fragmentary evidence that these apocrypha resolved the contradiction addressed in his *sotto voce* canonical pronouncement. It was this celluloid edifice that began to tremble, threatening to flake apart as his vision took on flesh and stepped forward, coming face to face with him at the edge of shadow formed where one of the slabs of light fell upon the floor, defining a weightless boundary across which he and she might remain separate though leaning close as if to speak, which in fact she did, addressing him in the voice of playback, a voice he recognized as the burial and reanimation of his own script and recitation, even as it gave rise to a clearly irresolvable misrecognition, crackling and hissing at him: *"My dear, my hand trembles so that I can scarcely write."* Then, turning a further half-rotation, so that he realized that the figure he had too quickly taken to be her face was in fact an elaborately embroidered cap worn sideways, she brought another, unexpected face forward to meet his, saying in an undertone whose subsonics interfered subtly with the rhythm of his circulation, "You had something like theology in mind, in which there'd be a song I'd have to sing that would turn out to be yours, or I'd be your song, or you and I would be words in a song settling in advance any trouble with the royalties, but I have to tell you now that the song that runs around and around in your head is one I wrote before you wrote it back to me for me to play it back to you, and that no part of me is part of that song, but that I am the component for which the custom stereo equipment wired into the walls and floor of this awful house has been waiting, so that there's nothing for you to applaud, just something for you to deal with, when you imagine me arriving on the scene, swaying softly in the grip of the chorus, singing, '*He's so fine, and he blows my mind.*'"

As if in agreement, the house did in fact resonate with her voice as she sang, making it absolutely clear to Birdy that, while he had tried to push the syntax in another direction, "My dear, my hand," was in fact all apposition and involved nothing like direct address, as the subsonics, subtle before, now jumped right off the meters, knocking the well-heeled crowd off its feet and flattening his own cheeks back around his ears, giving him the maniacal astronaut grin that, even in the years in which I came to know him, would at times lay hold of his face, as if the misplaced proof of a maternal prophecy that one day he might "stick like that."

"Oh, give it up, man, she's not even looking," muttered the first partial sounding above the fundamental of a family called Sodplug, as he pulled himself to his feet in the wake of the white noise of dramatic incident, picking up a stray contact lens that had fallen from his wildly rolling eye to the floor and replacing it against his cornea, dust and all. "And hey, your aura is seriously *fucked up,*" he went on, taking the wobble added to his vision by the bit of grit under the lens as a lingering remnant of bad energy in the air beside Birdy, perhaps trouble at work brought home and coddled, or the sort of conversation, at a too-intimate distance, in which some minor contagion, like last season's flu, is all that manages to pass between the interlocutors, and becomes the substance of their address, outlasting the conversation itself so that its memory, rather than the memory of fond words shared at leisure, becomes the jealously guarded object of the most intensely private investment. Of course it wasn't as innocent as all that, since this spectral component of a family called Sodplug had in mind the full harmonic edifice the name implied, a set of procedures sedimented in historical Enlightenment which would admit only a certain number of options for voice leading, and would thus require Birdy, astonished as he was at the flesh-and-blood emergence of she whom he until then had been led to take simply as an article of faith, to forget her face, to forego the demands implicit in her speech, to gather up all tokens of this recognition scene, except her name – for, although she had arrived on the scene out of thin air, in a nearly total breach of the composer's authority, still the right breast pocket of her auto mechanic's canvas jacket was embroidered with the name "Apostrophe" in exactly the spot where every other party attendant, Birdy included, bore a laminated nametag – and drop them one by one into the coin-op mechanism of an automatic response which, faced with anything genuinely unanticipated, began immediately an exhaustive series of researches into how best to arrange the furniture to make accommodations, which scholarship led inexorably into panel discussions and heated debates on the politics of interior design.

In the frenzy of activity that marked the emergence of this academy of the future, which was in fact a transposition of the academies of the recent past, made possible by their mutual aversion to primal history, it took Birdy a bit longer than was comfortable to notice that another instrument of a family called Sodplug had got his feet under him, and was busy tampering with the feet under Birdy himself – was, in fact, setting fire to his shoelaces with a disposable lighter. And if it took him a bit too long to notice the fire, it took him even longer, far too long, in fact, to translate that notice into action, sluggish as he always seemed to be in the wake of the constant waking and sleeping, waking and sleeping that divided his life into a series of impermeable episodes in an absolute and perfective past tense, though, truth be told, *from time to time the shadowy figure of a woman gliding up to him, murmuring a few words in his ear, asking him to take her home, would make him start.* All things considered, then, the Sodplug ploy worked like a charm, as the thought that some ill-defined other might be breathing the very air closest to him, burning it up in conspicuous consumption racing beyond all bounds, gave him pause and made him look immediately to his own arrangements. At this, a whole landscape came streaming through the blinds in sheets and planes, its unfamiliarity the strongest claim to this arrival being unbidden, as, with becalmed waters lapping gently at sandy beaches soaked in cocoa butter that went on beyond whatever distance one cared to look, and leveling the city into a flat dispersion of cabanas and piers crumbling into picturesque and fishy disarray, it rewired Birdy's nervous system to produce the subtly new perception that Apostrophe was not so much replacing all that he had previously thought to be the case, so much as she was, herself, out of place, a case of mistaken identity.

"Was that Esperanto?" he asked, smugly turning toward her what he must have thought to be a face as new in its own way as had been the surprising countenance with which she had first addressed him to the exclusion of all others, and in so doing had liquidated all the more thoroughly the appearance she had made just before of having made such an address while taking place entirely within an axiomatic he had, in fact, laid down in advance. While that revelation had been a matter of allowing the liquid color values of her features to spill beyond the figure they made, his present maneuver pitted fluid against fluid in the name of the solid state of shape, exuding the thick, biting odor of turpentine as it worked at thinning her to a kind of transparency through which, standing before her, he might allow himself to see the substrate just behind as a kind of dark and heavy outline coming forward to reassert the presence of enclosures with authority enough to harden feature into figure, figure into ideogram, ideogram into letter, as if this sequence represented the one and only form of legitimate history.

"His capacities are so limited," thought Apostrophe to herself, unfazed and unfaded, though thoroughly and astringently splashed, in exactly the space from which Birdy was in the process of barring himself, "that from this lexicon of spontaneous process, he has learned only to build a language of euphemism, in which to arrange the flowers around the conference table is to adjust the archive to ensure that I never shall have managed to exist, except in the supplemental vocabulary of commemoration. In such a manner, he and his patsy, this narrator infinitely willing to forgive himself for any shortcomings in his ability to imagine a real subject for his novel, hope to arrange the thread and trace of things through a thoroughgoing recursion and revision, and bring about a fairy tale in which my every appearance, over before it's begun, will have taken the form of *a pleasure which had never before existed, which they were striving now to create, a pleasure – as the special name they gave it was to certify – entirely individual and new*. For all this, their language – your language, whoever is thus addressed, the words you thought you were putting in my mouth when in fact you were simply shouting through curved air back into your own middle ear, loud enough to drown out your surroundings – is not even a palimpsest, but the complete destruction of its own every prior instance, so that there can be no accounting for what happens except as a series of disturbances in pure form, that is, disturbances that take place against no settled background that is itself the thing or scene disturbed, and thus leave me free to come or go, solid or wispy as I please, lyrically explicit or subliminal within your favorite tune, and not a thing for you to do about it but write it down in disavowal."

Dearest reader, it seems that someone has called and left a message for you to which I cannot bring myself to respond without distorting, not only the shape of the trust I presume you to feel for me, as I hope was made clear during my last appearance, but also the shape of the space itself in which that trust might find whatever could be qualified as place for it to take, and so I must remain silent, and by the same token can do nothing of the kind, since *other people as a rule mean so little to us that, when we have invested one of them with the power to cause us so much suffering or happiness, that person seems at once to belong to a different universe, is surrounded with poetry, makes of one's life a sort of stirring arena in which he or she will be more or less close to one.* So I'll leave the wrong notes for you to hit, and let your liberated dissonance unfold into a space exactly big enough to contain – some might say, "to confine," but never me, never me – that very dissonance, born thus upon the ground of its own freedom. Yes, thank you, your singing is precisely as bad as my score requires, and really, it's like sucking on a bottle of ether or, cheaper but almost as good, nitrous oxide, just listening to you go on like that in your straining head voice.

* * *

Once upon a time I could record a moment such as this, and listen to it over and over again at leisure without some hidden presence, some faulty wiring in the amp, introducing a buzz of distortion around the edges of the sound, seeming to hover on the verge of speech in some language closely related to my own and totally incomprehensible – but that was a long, long time ago, in the pre-digital era [see "Correspondence," note 10]. A similar moment, in which a young man named Birdy first discovered (or first remembered) that a silent, epochal change had occurred, was when certain immutable laws of physics seemed to enter into a momentary state of flux, some time after the dimly recollected event of a piece of musical theater that served itself to separate some other time from what had been, just then, the present. *And so he never asked himself what she might be doing, or what her life had been.* Occasionally, opening the refrigerator, he would find a leftover half-sandwich, bitten at the very edge by a set of teeth he might almost swear were not his own.

* * *

"Once upon a time," but that's always easier to say for someone like me, who makes no claim to having been there, than for a young man named Birdy, growing quickly older off in a cul-de-sac of his own, who nonetheless must have felt that time to be counted off "in his direction," so to speak, and thus positioned himself as the ultimate unreliable witness for the tales I have come to tell in which none of us really figure, him least of all, though he would of course deny this, and which are, for that reason if for no other, the very figure of "us" as we collected ourselves for publication in an earlier episode relating the course of more recent events, helping the form along, havoc along with habitation and simple habit.

"Yes...I never thought it would be like that."

Voices arrived as if from behind the air, and one had to decide whether to treat them as manifestations of a dimly intuited structure, a shadowy realm of scaffolding, rigging, and cables strung from girders high in the musty regions outside the light, and so to fear their sudden arrivals on the scene as a series of fatalities unpredictable on principle, or to grant them representative features within the landscape of one's daytime, standing not in shadow but producing their own gentle shade against the unwavering light, and providing something solid and upright against which to rest, with only the occasional disconcerting outbreak of ventriloquism to ripple the placid surface of illusion their rough permanence served otherwise to maintain. Which is to say, the scenery was always a step or two ahead of us, and even in those days of lower saturation the people stuffed into its hidden recesses formed a mass sufficient that its gravity alone argued the presence for which our senses could provide no evidence, and this argument created a language in retrospect, as history was even then progressing by its bad side, and so with language came the anthropomorphizing magic by which those features of our property, and thus those figures of our disinterested aesthetic practice which rhymed most closely with our own shape or scale, began to move around erratically, mouthing off until there were no secrets except those the trees themselves came forward to deliver. If I've told you once, I've told you a thousand times: my lips are sealed.

* * *

Once upon a time, a young man named Birdy sealed himself in a room, waiting for the décor to slip far enough out of the present to become fashionably historical, and so claimed this gesture for a politics of protest. In a tactical miscalculation, his reversal of figure and ground opened a crack in the subfloor, and Apostrophe emerged whole from the wreck of her own effects, a gas leaking into the house from its crawlspace, bringing address to the interior and placing the apartment number on the driftwood coffee table to eat through its veneer of hip indifference as if stripping off the polyurethane, so that Birdy found his gesture meaning everything he had hoped it might, diffusing through an entire atmosphere, but in precisely the wrong direction, cutting off his air supply.

"What appeals to me," Apostrophe intoned as if quoting from a source Birdy ought to know, and settling into a beanbag across the room from him, *"is not the practice of disinterestedness, but its vocabulary,* the way, especially, you seem to lack the words to speak that language without an interval of panicky translation, so that, to look at you, one would think you had a lazy eye, always squinting at some imagined tourist phrasebook out of the left side of your face, while your right trots out the repertoire of broad impressions to cover for its own partiality, face to face in earnest denial of any programmatic intent."

Birdy crossed his legs in conscious mimicry of Apostrophe's posture, figuring that, since literally everything had become a resonator and amplifier for her voice, and since it was the anechoic chamber of his own self-regard that had cut him off from participation in that global effect which had been his goal in conjuring her up in the first place – for so he still misremembered the scene of her emergence, calling himself demiurge when he was in fact something more akin to a ticket-taker – he would mobilize this diagonal symmetry of himself and the stranger across his rented room in a polemic against the lateral asymmetry of his own integrity, and would carry back as spoils from this battle a new body called consequence. That he called himself doing this was, however, the downfall of what was otherwise a brilliant plan, for if it was clear that such a plan depended finally on the cessation of belief, it was blindingly obvious in the connective tissues and integument of mutually retrofitted parts over which he granted himself survey rights in the present instance, that a belief had insinuated itself, and was in fact constitutive of that very fact of connection itself, namely, the belief in the absolute incomparability and infallibility of his own skepticism. And so he propped his eyes open with dirty forks from the kitchen, rigging up a system of flexi-straws so that he might weep directly into his own mouth and thus conserve this moment in which he could not remain, and of which he could not lose sight, so that in place of consequence he got a lexicon.

"It will become more and more of a habit," he promised the guest in whom he believed only slightly more than the promise itself, *"and I shall spend the rest of my life there."*

"Did it make your tongue feel good, then, that series of abominable mispronunciations?" asked Apostrophe from under a raised eyebrow. "Everyone knows where you hid the loot," she went on, pointing seemingly in all directions at once and making him tremble in anticipation of a final shame more surely than if she had simply pulled the laminated nametag with its official insignia out of his hip pocket and affixed it, along with the requisite lapel attached to the requisite blazer, to his shirt front. He read this feeling backwards and forwards in the only way he knew, and called his discovery a previously overlooked window in the wall of the hermetic efficiency apartment, through whose grime, so thick as to render the pane opaque in any but the harshest light of scrutiny, a man wearing a redevelopment agency badge squinted at him, scribbling furiously in a spiral bound notebook in script large and infantile enough that even from where he sat Birdy could make out the words, written out in all capitals at the top of a page, "AS IF YOU MATTERED," while the man interrupted his progress on whatever official document bore this ominous title to glance quickly backwards over his shoulder, as if expecting a sudden movement in, or of, the trees that clustered tightly around the one-story cinderblock duplex and its concrete parking slab.

"If that's what you want to see," Apostrophe mocked, sure that Birdy was incapable of understanding that most of what had been said and written prior to this moment was only addressed in smallest part to him, "why not look closer and tell me about the Intellectual in Red Bermuda Shorts, who dances around the Redevelopment Man's feet, just out of sight below the window sill from where you now sit, busily inventing a gestural vocabulary and administrative choreography for a dance called *Excuse Me, This Looks Like a Quality of Life Issue* – and tell me, too, while you're at it, why subject and object went so suddenly ambiguous at the end of that question?"

"Because," said a man in a dirty white lab coat, Dr. Duplex by name, who, sharing that name with the residential ground itself of Birdy's retreat from public exposure, found his expert knowledge and cut-rate consultations never entirely excluded from the scene, and hung around behind the half-height wall separating kitchenette from living room, "you are all so perfectly congruent, from where I stand, with the administrative situation you inhabit, that you are trapped in the ambivalent genitive – all except the Redevelopment Man, for whom the landscape is a bit of dative indirection."

"No one could be more fitly chosen as a patron by a secularized proletariat than that mother of a saint, to whom, incidentally, she gave a pretty rough time," piped an Intellectual in Red Bermuda Shorts, leaving Birdy unsure whether this was fact, opinion, or illustration by example of the grammar invoked by the doctor. And then, further complicating the tenor of things: "So it's utopian precisely in its untenability, and if you're not up to that rigor, I'll beat you to death in your sleep with a bar of soap rolled in a wet bath towel, you miserable son of a bitch. *Elevated...to the purple*, without the black and blue that might implicate me in the matter, you'll become a long-winded footnote by which I cut the corner between my publishing and your perishing. One...twelve!" he ended in triumph, further abridging, or merely setting back by an hour, the already truncated account of his time Birdy had managed to produce, or minded as one minds the office wall clock, and which he later passed as contraband into my hands, under the dinner table around which the family did long division on political minutiae in various stages of undress.

The door by now being blown entirely off its hinges, Birdy had a clear view of the itinerant muralist who, lacking a public school outbuilding or county health clinic to treat as a blank slate, was making preparations to ply her trade on the bare blocks of the place next door, and who, seeing that she was being watched, half-turned towards Birdy with an apologetic shrug, mumbling, "Anyway, it's faster than going door to door and spitting in every eye in the neighborhood."

"Oh, you artistic types are always arguing for the monumental scale of your own particulars," chided the good doctor, "but I'll tell you this much: I know a man, and several women, too, being always myself one or the other in my own spare time. *But of course it's all in the telling.*"

With that they all came crowding in at once, stuffing the bare room to overflowing, while the Redevelopment Man lobbed volume after volume of the municipal building code through the windows, until the sound of breaking glass was identical to his voice as he feigned agreement with all parties to the transaction, chuckling in a half-amused way, "Yes, what your realism needs is a vocabulary lesson. Here comes the dolled-up delegate of a family called Sodplug, late as usual from time travel back to that point at which it was last possible to stick his fingers in his ears and miss overhearing the sounds of demolition. Hey, how does it feel to be the last immovable object?"

"I'm sure I should be afraid; one can't be too careful in these episodes where, every appearance of discretion notwithstanding, one finds long division always going on in the forward-thinking background, and some remainder being carried like a grudge as whispers..."

[Tutti]: ...antiphonal in azure swing[9]

[9] [Eds.] The lead-in is a line from Hart Crane's *The Bridge.* Since this section of the manuscript is dated, it is possible to include here as well a quotation copied out in the author's notebook on the same day: "Space, I color it hard with blue till it almost melts" (Tisa Bryant, *Zoo Kid*). The connection between this citation and the opening of the aria is obvious, though it is unclear what either might have to do with the language of the song itself – which is drawn primarily from Adorno's *Minima Moralia,* various philosophical papers of C.S. Peirce, and Melville's *Billy Budd.*

[Sodplug solo]:

...but for just this reason constitutional
precipitate finds indeed the fatal
rectilinear victory embraced by this
dynamic. Which is quite another word.

The child feels herself. Mysterious
leagues of grownups make a feast-day come
to sanctify it. Come close and fast from distance
measured off in slow archaic leagues. The child
feeds himself. His appearing then: nothing.

The child phases itself, feeds
back. Yourself is speaker noise

like psychoanalysis in the most developed
country or product that listens at all,
denies its deviant subject matter
in advance. Even trial is becoming
merely biding time.

"On the *grow*"...
few people care.
Already I observe
that this, that other.

We come to full possession of
our drawing, one reason or
another — but whatever.

Glass shattered history
or science, according

to which very small changes
have not yet taken place.

Reproduction polychrome

reflects you present, to whose doubt
those features represented categories.
I am sleepy, and the oozy weeds about
me twist. But still I have this *interest.*
To reasoning, truths which *must be something*
stand to someone for capacitor
and rheostat. Address that feeble
thinking of some prior thought,

and bring forth one more lollipop[10].

[10] Once upon a time, you could say anything nice, but you didn't say anything at all. Diacritical ultimatums interrupted dinner with definitions withheld from view so as to grow huge and ravenous in the dark. I am making this up, of course, as others made it up then, and thus the accusation leveled by a doctor and anatomist of fact named Duplex, had it been voiced at that time, would resonate still and call me to account, but since after all I *am* the one who makes this up, and since I did, and since what is done is done in that image, it turned out that the good doctor had mortified his voice as an act of contrition for the grossness of a speech impediment as of that evening still entirely prospective, and so no accusation was made, though a list of sorts was made up. Have I proven my spontaneity to you, and do you feel free of the memory of that night, reading over the transcription? Have you swallowed your tongue, or mine, and can I make it up to you?

THE ENSEMBLE EXISTS [11]

The Ensemble Exists
beginning microcosm
—Cecil Taylor, "—Aqoueh R-Oyo,"
liner notes to Air Above Mountains

it would have to be lost so thoroughly as to breathe
— Leslie Scalapino, New Time

You say you have this
talent for the said.
To rest in that is
true — as in default
of sense beneath whose
crust run faults, cracks
full of serious,
obvious wet.

The habit of that subject is the plan
of a beyond which values a succession.
However all and each may speak of chance
employment, every character is one
and only one. Two or more at most. Here
we have a logic of the subject very
frequently violated. Our one man

[11] [Eds.] Are this poem and the one that follows to be taken as arias like the preceding? Some formal markers would seem to indicate such a reading, but others are missing. Furthermore, the large caps of the title line correspond to the style used to indicate chapter heads consistently throughout the rest of the manuscript.

(or two or more) for the occasion,
Dr. Lyon Playfair, for example,
takes gravity to specify each toe
on which he stands alert, a point *en pointe*.

Hold that representation. Admit only
the material as in reasoning only
to the real, as it really is. And there it is,
belief in metaphysical fictions,

which "man" means there must be
beneath this thing
incited contour[12].

[12] Once upon a time a cloud of sobbing strings pretended to protect the figure of a woman in what was either a song or a landscape, hovering above her and around her like moonlight and water, but only in the sense that moonlight serves as the medium for speculation on nuclear fission ninety million miles away in the opposite direction, and water acts as a lens, focusing what it covers into a disclosure adding the picturesque to the particular. There had been an episode of some kind, starvation had loomed, she had been forced, not without some ironic pleasure, it's true, to eat the still-squirming flesh of an Intellectual in Red Bermuda Shorts, among other things. Some of these things were clear, while others remained titular. Under the titles were texts, lyrics she was engaged to write so that a man named Birdy, no longer young at all, but who saw himself now as he had been ten or fifteen years ago, solitary and intransigent and correct in a concrete box that housed a single room, plus toilet and kitchenette, might within those texts produce disengagements by which the letters would emerge as black designs against white paper, and behind them, whether reading or hearing them sung accompanied by guitar, the people of this landscape might then intuit the presence of himself and his Apostrophe – for thus, tropically, did he continue to think of her – she beautiful and he evil, or vice versa, or all one way or the other, so long as there was some sense of distribution implicit in their roles. He identified the luxury sedan in the commercial whose jingle made him think of her, and bought it with the last of his royalties from a previously pilfered tune, and named it As Such, driving it around everywhere, even on rainy indoor days, so that the city and indeed the nation might fondle the exciting and aerodynamic contour of everything that was surging within him precisely as it occurred. *And the pleasure of being a lover, of living by love alone, was enhanced in his eyes, as a dilettante of intangible sensations, by the*

Or too late, if semiosis

will not be which sign is used.
Now such hypothesis seems
any number of reasons to be
recommended to the foldwise
irritation of the fissile ones
and twos who would pursue my
fondness for a measured stroll.

I study the varieties
of humanity that move me
and still retain the element.
Even a least, peripheral
sensation of pleasure, of
pressure, tongue sliding over
fissure, has its excitation
and molecular gridlock.

The principle with which one sets
this weight upon a true continuous
plumb, a line of chalked descent,

is named without appeal,
trafficking in outlines
on the bare, black asphalt,
just inside the *simply there.*

price he was paying for it *[see "Correspondence," note 11]*

Apostrophe, watching it all in slo-mo, said to the unidentified others sitting next to her on the dilapidated couch in the television's blue-tinged glare while the cheap VCR hummed softly, "Here's the part where he begins to think he's outsmarting – who? not me, even he must have given up anything but the broadest of analogies in *that* direction – but someone, nonetheless, who turns her head from side to side, allowing his right hand to drip something oily and bitter into the balancing mechanisms of her inner ear, while his left buys out her lease. *It becomes possible to cut off the supply of light from a house.* Look out for the landlord who tells you he sees through the vanity of possession, for he's about to knock holes in your walls through which later to knock a hole or two in you."

THE ONSET EXITS AND THE DAMAGE DONE

...to treat philosophy itself as impossible.
— Louis Althusser, For Marx

One of these good old days, I'm gonna call your name.
— Al Green

Though made however visible, the signal
success becomes an end for putting up
one's evidence. The result is what sagged
thick beneath the contrary pole, sort of handed
down from the dishonor attaching to
productive early consequences of
your ownership. And so it is to be
"community" no longer but a stain,
dependent for its livelihood on yes.

Form of wasted effort
Apparatus evident
In evidence rendered
Leisure in "housewifing"
Rendered as a verb

to some few cheap false goods conspicuously
outset in philosophy. The floorplan,
the voyage to the backing of the mirror
and back to front the openness of this

enclosure [13]. Its cross-eyed binocular
windowing of options from the desktop
horizoning the line of sight behind

[13] Once upon a time, you were expected here, and when you didn't show, a place was set for someone who looked more or less unlike you. You are sure to congratulate yourself for having caught me in this situation, making it up as I go along, but no matter, my aim's to bend all light into reflection, and so I have pre-empted your discovery of this point of fact. *In this strange phase of love the personality of another person becomes so enlarged, so deepened, that the curiosity which you now feel stirring inside you with regard to the smallest details of my daily life, is the same thirst for knowledge with which* I *once studied history.* You have knocked a hole in someone's wall and now there is nothing to do but look through, even as you realize that here is not the character for whom you looked, but perhaps a near relation, a time perhaps a bit early for your arrival, in which a beakless old waterbird wanders in circles over a threadbare carpet, casting terrifying shadows out onto the sidewalk as its circuit passes and repasses the lamp to which one of its webbed feet is tied by a length of packing twine that shortens inexorably with each revolution. Perhaps it has been a long time since you made any further holes, any perforations in the thick-napped paper that covers the interior walls of such a scene, simulating an all-encompassing sky which wraps itself around the outside of a train of events you were only moments ago riding as a casually interested commuter toward some other destination entirely – a long time, then, since you were a child, shooting at every hint of movement with a pellet gun by the shore of the artificial lake behind your father's rental duplex, knocking holes in beaks of birds to create the asymmetry out of which a story might take flight, *but in this case the mind, merely by recalling the pain, created it afresh,* lifting "that of which we cannot speak" into a chiaroscuro of yellow lamplight swirling down a drain into darkness as warm showers fell upon you from the hidden nozzle of theatrical night.

"Tell me everything you saw, and what you think it means," stage-whispers Grace Kelly in the fascinated hush that cuts off the supply of light, her absent presence in this vignette responding to your recitation, which takes up each sentence in the discarded papers blowing around against the all-encompassing sky and makes of it an opening whose dimensions are given in the aperture of the vocative "O." Framed thus in the round zoom lens of surveillance, do you now see the dashing figure you cut from afar brought into the tightest close-up, made ugly and atomic?

the head.
The drudged
caption floats
to title

justified columns that
direct each direct
expression of classed
schemata, bent space.
Past life replanted
in the yard that
hasn't time for second
planted feet to take
the straddled upright measure
of ruled, unruly stance.

A certain number bend their energies
to shape, living up the walls. *Time enough*
for severance — give that repetitive
stress injury a final grunt by way
of compensation. When you think
to leave, your key drops through the slot, like mail,
stamps you with the ring of self-address.

* * *

Once upon a time, there was a storyteller who introduced into a tale *one of those fragments of literal truth which liars, caught off guard, console themselves by introducing into the composition of the falsehood which they have to invent, thinking that it can be safely incorporated and will lend the whole story an air of verisimilitude.* Apostrophe runs away, as geological layers of forwarding addresses turn the multiply-enveloped scene of the narrative into a palimpsest, with reference to which the only certainty is that someone will receive a paper cut rather than useful knowledge. However, in another tense her face pressed out from behind the paper as from an interior, all velvet-papered walls and lamplight, which the storyteller thought to foreclose every time the story drew the curtain on a scene, the contour of nose and cheekbone solidifying for a second in the soft billows of that gauzy fabric which served only to insure that an audience's interest remained with what the tale hoped to leave behind, in its jerky gait stumbling from episode to episode.

Again, though, I am lying, as she draws the curtain back and lets the light fall upon you there, a bit below her in the street outside, where all presuppositions stand together in first and second persons, and says, "I am not a bad habit, not on the order of the forces 'driving the urge to differentiate, to identify one and then two,' only for you to make the subsequent half-turn into saying that the latter follows the former, as misdirected speech follows curiosity about a lighted window in a cinderblock rental duplex on a dark night with all the sodium vapor lamps shattered, duping an accomplice into accompanying you to that theatrical "bad neighborhood" in which the passing audience method-acts its way into belief in the constitutive powers of its own direct address to the emptiness of its relation to the stage properties, when in fact I have been here, and have been telling the story of this circle of light, all along."

* * *

Once upon a time, there was a man named Birdy, older then than now, whose job was to remind a text on memory of where it had been and why it had missed itself there, as if he were a third-person argument between a first and a second person who tried to keep appointments, but whose lives, in the event, were compromised by one emergency after another, so that one or the other always got stuck waiting, picking up the check, taking the wrong bus home.

"Otherwise," he said, coming to his senses already in an anaphoric mood on the curled linoleum floor, decorated in a grease-beclouded floral pattern, of a dimly familiar kitchenette, talking to no one in particular, or else to an offstage presence stuck in an abandoned episode, "you'd simply live the singer-songwriter myth, and have nothing to say for yourself but a lot of florid weeping and moaning. I knocked a little hole in your wall because I care. Pornography drives the photograph forward into full relation to its time, and becomes a method to produce clear prospects on what was once obscure. I continue to love your body, the taste of your tongue and sweaty ass, only because I simply can't imagine you, and see not you, but a broad expanse of sky, on the far side of every window into which I peep."

* * *

Once upon a time there was a man named Birdy, somewhat younger than he had been in an episode just a moment previous, who gave the lie to epistemology and remembered the exact contour of every pain his body had ever suffered through, beaten into shape, even, perhaps especially, when his mind had been on something else. Sitting in the dark and hearing, from a door or two down the block, soft blowing on soprano sax, he experienced someone else's clarity out of phase.

"I was afraid he was going to eat me," a particularly disingenuous character in the drama of a family called Sodplug was just then reciting, in rehearsal with the cop with whom, arm in arm, he was coming up the walkway of half-buried railroad ties that kept the path from parking slab to painted steel front door relatively dry and passable during the summer floods. Of course, since there was no air conditioner, Birdy had propped the window open earlier in the evening with a stack of crank pamphlet literature, and so heard every word, but refused on principle to believe in the reality of any of what was said under such circumstances, and was maintained in his condition of exemplary, self-consuming skepticism. By the same token, though his skull was perforated as proof of the most certain kind, he could not grant the veracity of reports that told of a mutant, cannibal freak lying just beneath the surface of the duplex and chewing on the sleeper's head during the sticky night, but maintained that Mr. Article was, if anything, a fiction of grammar granted temporary substance by the growing thickness of ideological descriptions of landscape. Meanwhile the sax was licking lascivious promises into his ear from distant, microscopic horizons, so that, when his one hand found his other hand scrabbling around his lap in the dark, his first impulse was to turn on the overhead bulb and

find a window envelope, write a check, and mail away his first adjusted payment.

* * *

Once upon a time, my little bird, you were saying. Then the therapeutic architecture, its medical practice sprayed thick over walls and fixtures alike in anticipation of each new tenant, opened wide in a tiny, repeated gasp of cancellation, swallowing your evidence in this incredulous and barely articulate displacement of pitch and timbre by pure, iterated rhythm, as if it were a quiet conflict approaching silence, becoming a branded product, and we promptly forgot it all, and had to wait to recall you as an after-effect of our own belated recognition of a housing crisis.

* * *

Once upon a time there was music, the ratcheting clicktrack of the manure spreader we landscapers used as a backbeat to drive the more sophisticated layers of performance along, laying down a thick groove or furrow as the medium in which to scratch out together the tale of our growth in alternating cycles where shared biorhythms, even communal organs, would give way again and again to the paranoid erection of survey markers and bounded regions.

"You are a formless water that will trickle down any slope that offers itself," I find myself now having occasion to sneer, in this dream I am busy having, in which I am a man named Birdy, older than the younger self who will one day come to receive of me the wisdom I have extracted from my episodic encounters with others who always were, as the old saw has it, myself as well, *"a fish devoid of memory, incapable of thought, which all its life long in its aquarium will continue to dash itself a hundred times a day against the glass wall, always mistaking it for water."* But when the object of your insult is by definition elsewhere, apostrophe grows stronger, the minuscule approaching the majuscule with every such attempt to inculcate a particular reception. I'm not thinking very clearly at the moment, my vision blurring in anticipation of its double, but it seems to me that I might spend some time here looking for you, even though I'd be unable to stand the sight of you were you

simply to appear, and so I have arranged our introduction to each other as the exchange of pre-erased portraits, layers of commentary on missing objects, "transmissibility without truth value."

Contracted at seams that swell
armatures (frames), arms embraced re-entry's fire.
To arc cold lumps from ballast, to tabulate
so under brackish tongues, licking

signage. While you're out, its aggression might
imply just atoms. A stake,
and the heart fled the matter.
If fact's ass-backward, body'll drag,

an issuer of interminable elimination.
States border ends. Congested tissue's
ample, and just meets a stake in
royalties, singing at suggestion while it might

yet turn tooth's weight to worth.
Bolt into culpable works—away with you.

Your faces – a collective and formless mass – escape the grasp of my imagination.

Or put it this way: a naked emperor, or presidential imposter, or a real president, or Henry Kissinger, was covering vast tracts of landscape in rancid shit, and it turned out that nothing got you quite so tingly as to strip down and roll around in all that hot, poisonous fertility. I am allowed to say this because others have said it before me, though, it's true, usually with myself in the prone, besmirched position into which my defensive hiptoss has thrown you. I have taken to following your car around to all your familiar haunts, my hand trembling so I can barely write, or steer. I must be very, very clean, wrapped in the sacrificial rayon shirt and tight, faded denim in which I am invested by way of my placing you at the vanishing point of every scene, where you hold all the threads, as it were, and are thus invested with the power to weave the dim fuzz of participation into a halo of radiant natural and synthetic fabrics upon which, were you once to wipe your hands, the stain so placed would dissolve the whole ensemble in reaction with its obdurately proto-biological proteins.

I can hear you, one after another, a long way off, coming nearer, passing my door without stopping, and bearing away into the distance a message which is not for me. But your faces – a collective and formless mass – escape the grasp of my imagination, and fail to feed the flame of my jealousy. So I tell myself, until I catch you in the banal and predictable lie that you were, at just that moment, passing through a police checkpoint just outside downtown St. Petersburg (Florida, not Russia – it being after the time in which the latter St. Petersburg had existed, and before the time in which it was made to exist again in what was claimed to be its former state), stopped and questioned as a potential rioter or allowed to pass with a warning and a grim joke that lifts a corner of the draperies from racist state terror according to a practice not yet called profiling, so that I must imagine, head-on, a face for you, and attach to it all the longings my belief in the at-will nature of your employment among the travel agents and other staff of my sentimental journey has held in abeyance, if only to verify that such a face, in such a situation, could not have been your own, and that in fact you could not imagine such a thing for yourself, even on your own terms, having done more than enough to imagine terms for my imagination of what it would mean for you to be instead reclining in the cocktail lounge of the international terminal at some airport or other, sipping an overpriced beer through different lips altogether but now irrevocably if indifferently lipped, eyed, *faced*, singular, dear. All of which is a past-tense kind of loneliness, save the checkpoints, which persist in one sociable occasion after another as the reminder that the bottle from which you sip will be presumed to contain various liquids under pressure, like that pricey beer, for example, shaken, in place of the prospect of a child yet to arrive among the more recent tenses of a previous chapter, into effervescent opening, so that you decant propulsive force creating distances across which a descant of our mutual absence might emerge, the child of a ballad growing more revocable in time as its grownup analogues grow less retrievable in space, a church or civic center reverb for a song one or both of us might sing about what else the bottle might have held, as the silence of the wide focus crowd perspective on the evening news takes its shape from the silent "e" turning "sing" to "singe," which is the rush of air into, rather than away from the bottleneck, flames like signals by which that "collective and formless mass" writes a series of its incognitos in scorch marks on the sides of police cruisers and the steps of city hall, where the lingering after-vision of the segregationist mural flash-burns itself into memory as if the contents of that building, burning itself out, would always disclose themselves again for eyes closed against the heat, leaking fluids, fighting fire with fire. This might be a bit of real life I am inventing for you here, alongside the alternate ending, a prelude in which you jumped out of your skin and ran down the road [see "Correspondence," note 12].

In absorbing my dreams, you have delivered me from them. Examining my complaint with as much scientific detachment as if I had inoculated myself with it in order to study its effects, I tell myself that, when I am cured of it, what you might or might not do will be a matter of indifference to me. And if I insist on the itemized account until then – well, call it comparison shopping. I have spent the night memorizing lines and, as you might well know, one line equals one answer, a line of erasure for a line in pencil, rubbing out stroke by stroke the blind contour drawing that was my first representation of you, replacing its incised black with an abraded whiteness whose margins are decorated, as by funereal cairns, with small piles of paper fibers rolled around the bunched rubber of eaten centers, law of conservation of energy up against law of hemispheric hegemony, while even the word, "voodoo," that promptly springs to hand by means of locating this simulacral violence, inaccurate, deflected and distorted though it may be as description, retains its precision as a reservoir within the arid language I have hoped to construct by building outward from the page on which I draw you as a blank, from which I draw a blank as you, this one misrecognized word serving as the reminder of an inaccessible depth revealing nothing but further reminders of persistent reawakening to the way even projection – of halogen light onto a darkened street full of badly-photographed masks in mass movement, for example – borrows quantum substance, in however perverse fashion, from exactly those bodies it disavows in laying claim to its own unobstructed, nuclear ability to come before every eccentric orbit, to embody the plenum prior to the scattering. But go ahead, ask me anything, except the one about why a man with a Geiger counter meets you in the foyer of your apartment building every morning, which will answer itself when *the face of another Apostrophe will appear and rise before me, softly radiant,* certifying the payoff of my ugly indulgence in politics with the halo of a bona fide ethical experience.

On the 5 Fulton this morning, just so you know, everything was wrong, not least the awkward intrusion of present-tense San Francisco as a translucent skin or acetate through which a past-tense Tampa and St. Pete, an odd mirror of a "Bay Area," was still projected, and also the sense that all eyes forward came careening back at me as if a rearview hollowed out recesses in the foreground, while the driver met me head-on in the middle of a rush of acceleration, the slick road chewing up the bus. There was a small group of young men huddled together with a boombox in the last row of seats, coming in at odd moments in a kind of staggered unison with a track called "Playback," in which a lost plot with which I once had been involved was found again to be hatched against me, a head-gnawing crunch of a rhythm section seeming to come from the ground itself, as if the bus transmitted an anachronistic chewing upward through its seats as the means of individuating, attaching articles to its passengers. Thus, too soon, ended an episode I meant to call "The Music Lesson."

Playing back that prelude, the learning of whose passages had been a bit like reading the off-hours column on a transit map, lateness intensified into evening as a dawning of the past tense, balanced on my recognition of the cusp upon which "late" itself teetered, on the one hand a present so far along as to spill into futurity, on the other hand a past so far gone as to be its own memorial inscription, roosted upon which cusp, alternately staggering, swaggering, and swaying, I found Birdy once again, a bird on a wire reading the telegraphic import of the moments into which he had traveled to prepare the scene for my own consciousness of the drama of recognition we were slowly playing out, paying it out like slack, paying it off in small installments to keep it quiet.

"Don't worry," he whispered in my ear, sliding into place behind me and deftly finding the hidden mechanisms by which to move my mouth, transitively lip-synching me by not-quite-remote control to a song whose sounding remained just out of reach, which I recalled into the air of memory without having first addressed with the fluids of my tongue and throat, so that he might intuit its presence preceding my every inarticulate and arid sigh, even as I failed to make the slightest mention of it, "no, don't worry at all, I've got your back." So there is an ulterior motive in the more direct approach with which I have turned my face in your direction, dearest reader, as behind my every word are those words of his, uttered in a situation in which to think of words was from the first to think the better of them, by which he promised to stand by his apostrophe, if only to stand by while Apostrophe lifted off from his lips and took flight into atmosphere, so that by this multiply-convulsed and inverted ventriloquism he might maintain surveillance, reception, radio contact with the absence to which he'd applied his signature as an *envoi,* all the while staying deviously true to the spirit of that contract according to whose terms the space into which those words escaped would contract expansion as a disease which ran the Zeno's paradox of its course in asymptotic approach to something final and fatal. *For a period of abstinence may be calculated, by anticipation, as having lasted already until the final date assigned to it,* and the entire use of telling stories such as this in the past tense is the shiver of cold delight they give us as we peep around the corner to the deaths they foretell, the death they forestall, and the forests all chainsawed down to nubs and stubble save the strip of lush evergreen running parallel to either side of the highway, three trees deep and stretching the thousands of miles back and forth from Florida, say, to California, or at least another, more exotic part of Florida itself. I could have kicked the habit in that town, or that one, you remark to yourself, highlighting the dots named "Frostproof," "Lorida," or "Indiantown" with pushpins, perforating the flesh of the terrain along the creased route which marks the edge between the homeopathic self-similar and the mnemonic substitute. But for Birdy, as for me, the one this address hoped to locate, grounded or wounded at the right coordinates, had nonetheless a different shape entirely from the incision he or she had made in the soft tissue just beneath our tough shell of outward reserve, notwithstanding the attempts, whose retelling by now has surely begun to tire even the most assiduous reader, of a certain articled individual identified with the subcutaneous layers of the landscape itself to make a distinct, articulate and lasting impression directly on the contour of our now-shared, now-apportioned brain, an organ for which we ripped a page from a passing drama and, with a slight revision in the spirit that seeks to reconcile right feeling with right counting, coined the name Two Gray Sisters More or Less.

The echo of this tentative doubling is the ineradicable cyst in my right earlobe, throbbing with sense-memory of events that have occurred to me only as hearsay. Was this the same sort of memory in which was grounded Birdy's automatism as he phoned in his order with Tampa's first delivery-service Szechuan restaurant, just as he phoned in so many of his performances, not wanting to leave the zone marked off by his four cinderblock walls, within which was a fairly small margin for error and stage for action in his echo-location of himself, even as this predictability and constant predication were founded on an essential dislocation, both his own and others', that allowed him nonetheless to "hold the line" for the ring of his beloved Apostrophe, an instance of that predication, unpredictable with the static-electric burst of prediction itself, that motivated a "sounds like" stripped of self-similarity, a conduit for which he held the negative pole of a blank and always ready likeness, waiting all the while for the positive access and excess that would startle him as looked-for but unheard-of sound? By which I am reminded (though why I find my own memory returning in such clear outline at precisely this moment must remain a troubling question for a later moment still), that in a later moment earlier than the moment just now parenthetically named in those terms, "our own" Americana first acquired the habit, or rather the costume, of ghostliness, arraying herself in a heavy white comforter, at first only on Halloween but each year spreading out a little farther in either direction, getting at last well into summer and winter, muffled by the thick layer of down from acoustic contact with a world of traffic at dangerous crossings and hoping by this disguise that functioned in reverse, that is, functioned to place her more "in the way of the world," to enter death at the crossroads having already become a spectral presence, in other words to haunt death from within. Something in this scenario calls out ahead of its time for the invention of the unanswered answering machine, the call screening through whose mesh Birdy would ideally have felt his attempts at contact sieved, their halting, apologetic and self-lacerating rhythms anticipating the feel of talking to a tape in the way they placed him as both inarticulate speaker, rattling and stammering into a tangle of wires without seeming outlet, and unappreciative audience, listening with a critic's reflexes from the position of the other to whom his colossal presumption of speech failed to extend. Just there, in the unanticipated roommate's position diagonally across the living room – so called because it was the single room in which all this living was being done – my Uncle Madame exhaled a cloud of resinous smoke and sang out, in birdlike voice that borrowed and failed to return the nametag of the one-sided conversation for which he served as wiretap, the sweet tritone of disconnection, *diabolus in musica* given contour by a cheerful and devil-may-care grin of sheer good-buddy camaraderie.

"Look close," he chirped, "and you'll see my fingerprints on her. The whorl pattern in the center of the right thumb, drawing the lover's gaze down to meet its own demise, will serve, years from now, as the perfect type of the *impotent, blind, dizzy anguish in which those early years of Nixon's presidency had been engulfed,* or so you'll tell the chronicler, having found a language for your defeats more precise and coherent by several orders of magnitude than the dim future perfect into which you stamp your imperfect impressions of utopia." Upon recognizing this admonitory tune, Birdy smiled and relaxed into the dim buzz of participation which was its B-side, preferred by the subtle collector to the more populist appeal of the single's pop hooks for its self-abnegating hold on atmosphere, mirroring as on the outside surfaces of reflective sunglasses worn perpetually against the semitropical sun the sense he had of himself as absolutely crucial to this diffusion of Apostrophe into the general humid haze, positioned with literal precision at the crux of transformation and in that sense invisible and immaterial to its history the more his secret sense of conspiratorial insight into its workings intensified.

In other words, Birdy found himself hard at work throughout the better part of the next decade on a piece of junk sculpture, conceived at a monumental scale and ultimately to remain unfinished, or else consumed in conspiratorial fire for the precious cargo of unburned petroleum with which much of its salvage was liberally coated, a conflagration so huge as to flavor the surrounding air for miles around with an acrid and eye-watering tang, and yet so tenuous in its hold on the status of "event" as to merit an almost complete invisibility in the guidebooks, mirroring in this paradox the major occurrences of the decade yet to follow, during which Birdy mourned the passing of his masterwork in the gradual accretion, one word per year, of the posthumous title, "Ten Thousand Foot-pounds of History Compact Your Used Sedan, Citizen." My Uncle Madame kept an eye on the inverse sky of this toxic ziggurat from within. After the fire, reconstructive surgery made his face look like Birdy's, only more so, and his report on what had transpired in the world during the years of his captivity captivated his new twin with its resemblance to a series of experiences he was already most willing to allow himself to have had. Time runs in strange eddies through that choose-your-own-adventure porn, in which the heads have been masked out and hand mirrors are made available in the theater lobby – and, more recently, velcroed to the back of the videocassette case – so that Birdy's nudity became the conventional shorthand for Uncle Madame's face, which is still the reflection I confront when playing dummy, the reserve of speech I tease you with when I play dumb, turning over to present you with the rump of possibilities gone stale in sheets long unchanged.

And yet we should like to live until the time comes when we no longer love you. Until then, we savor every nuance of the reports that have you plowing your used sedan through the checkpoints, then coming back for more. From where we stand, or rather stoop, our heads inside each other's clothes, perhaps we do idealize a bit your capacity to do damage, and perhaps this is the beginning of a disagreement between Birdy and myself (Uncle Madame reserves the right to speak for himself), in that I'm not at all sure you should thank us for attending to the devastating cut and stitching of your radical leather pants, rather than the devastated lacerations and last-minute stitches that might constitute the scene of your engagement at the barricades told by other means, especially since, if I read the constant whispering in my ear correctly, we'd envy equally the pair of hands that decked you out in style, or laid you out in state, all such romance of contact being equal for those who won't be bothered to imagine more precise terms for it, the low-density cinderblock "slurb" being precisely matched to

the heart of the town
as it calmly
and quietly slackens
into airs called simply "air," or "atmosphere"
The destinies morning had decreed became a state somewhere
It's "provisional self-government" when you think
to pet the disarranged
transactions of your delicate head   terrifying
minima distinguish chalkbox from nightstick
...subsistence what I think they thought of us   — decide to get
in place of foul air and track lighting
punched out hourly by the clock
and disappeared. *A skeleton for revenge.*
Idealist bullshit   ...we burn a few leaves out back
occur at finite distance from
the kid they hogtied and then choked   ...memorial
in blinds, in double-blinds, in every survey.
Scratch choruses return to fire

the broken cookie warehouse outlet
— a broader, slower arson lights the bodies up

without examining the same
The child feels herself mysterious
as the constant traffic rhyming neighborhood with vacancy
[see "Correspondence," note 13]

in back. Yourself is speaker noise out front,
advanced. Every trial is becoming
their notation, temperamental music of exchange
— like, food stamp for green stamp — like counterpoint
he pulls taut now, sore sum
less precipice than pothole
Reproduction polychrome
Turned the prayer wheel through my own
power needs. Two-hundred-dollar heat

and a talent for the said
Had learned and forgot
The habit of that subject is the plan
...duct tape on my shoes a birthmark
like his tiny body like
take gravity to specify each toe
or finger lost in counting ten... nine...
— of what she'd keep if they forgot her in the distribution
...eight kids clustered in the drainage culvert
Need accorded to bizarre collections...seven...six...describe
the standard-issue shoe that kicked your ass, five toes behind
that steel and still this fondness for a measured stroll
or music that I need when dreaming what
we lick isn't only sidewalk
but each other, plumb and foursquare, a line of chalked descent
no eye has eyes to understand

the day you had ...and where the paycheck went
a pistol followed
— "community" no longer but a stain,
the bricks that broke him open, with some help
The home that sold her body to itself reflected
voyage to the backing of the mirror
...piecework typing taken in a saturated screen —
accommodate this promiscuous image
What we circled What
past life replanted, how many days ago
...three...two—yes, but I don't need those flowers
in condolence when a single city
or product, listening in, would be all of worker's
compensation, once you stop the count to think

"Well, now that *that's* out of the picture, we can get back to killing time," sneered Uncle Madame, his gesture mirroring exactly that brush of the arm he would later come through force of habit to make over and over again whenever I came too close to the picture plane of the television, as if it were a windshield one had to watch constantly for the threat of oncoming traffic, lest the big picture it kept bringing in from far horizons shatter into fragments that stuck in one's face. "Isn't my body ever so much like a landscape, a whole system of resources linked and crosscut by a web of highways making provision for beauty, danger, even dinner?" he asked, pinching a bit of yesterday's fried chicken skin, shaped curiously like the old war criminal just then addressing a press conference on the flickering screen, from where it had fallen into the fold between shirt bottom and waistband. Invisible assistants stood by, thronging the periphery of Birdy's vision as he glared impassively at this latest bit of provocation from his roommate, waiting for the sign that would send them to their surgical work of certifying that ungainly body clear of all mitigating hangers-on, the bits of meat that meant the world from which they had been gnawed.

But even this ugliness of faces which of course were mostly familiar to him seemed something new now that their features – instead of being to him symbols of practical utility in the identification of this or that person who until then had represented merely so many pleasures to be pursued, boredoms to be avoided, or courtesies to be acknowledged – rested in the autonomy of their lines, measurable by aesthetic coordinates alone.

"Oh, fuck me, now we're back to the whole abstract line drawing business," he sighed to himself, "when all I was ever hoping for was some really baroque, gratuitous sense of splatter, some sucking chest wound of a face that would, literally, suck the breath from my chest along with the articulations from my tongue, and serve to confirm me in my phantasmagoria of feminized horror – some sort of externalized male womb, or that 'monstrosity of male love affairs' about which Genet writes, and which always gave us straight guys such a titillating frisson every time we got exactly that far in *Querelle* and then put it down. Instead all I get is this crowd of checkout clerks, armed with mops and armored with bits of broken windshield prickling around their eye sockets, each one a fish *carrying about on his person only an accidental and perhaps purely symbolical fragment of the glass wall of his aquarium, a part intended to suggest the whole,* the hole from whose real appearance a recoil into the 'purely symbolical' was precisely my point all along."

The music meanwhile soared along above the clouds, at odd intervals pushing one of its rapt captives from the cargo hatch and describing, with the barest rhythmic hitch, his or her screaming descent to the packed plain below, as if a bit of bottom had dropped out of the rhythm section, reminiscent of early recordings of New Orleans jazz in which the rather more stable continuo of breath through the tuba had to stand in the gaps through which the wax disc would have let slip the percussive attack of a finger, pizzicato on a contrabass string. In much the same manner, I have subtitled the film during this sequence to compensate for Birdy's unselfconscious regionalism putting over on itself the pretense of universal speech.

"OH DEAR ME I HAVE LOST MY HEAD WHAT A JOHN-THE-BAPTIST MOMENT."

[Many frames of the moon follow, shot in different phases, from different angles, on different film stocks, etc.]

"WHAT A CROWD IT MUST BE OPENING NIGHT THERE SEEM TO BE CERTAIN BUZZWORDS REPEATED ON ALL LIPS."

[Birdy dances a mock tango with his own reflected image in the shard of safety glass lodged in the corner of a men's-room attendant's right eye.]

“IT CERTAINLY IS A CATCHY TUNE I CERTAINLY HOPE THAT I DO NOT CATCH IT.”

[But Apostrophe's method, her beautiful tone, came to life now and then beneath the fingers of her pupils, even of those who had become in other respects quite mediocre.]

Thus it was that Birdy swooned to his senses and realized that he now stood in a cinderblock box of a room, with not a mirror in sight, just as he felt most acutely the need to see what his lips were doing, literally to watch his mouth, whose stream of extruded plastic instability manufactured substrate and housing for the hardwiring over which the code of days to come would flow. And thus it was, in turn, that he came into the curiously bent posture in which we found him in my own childhood recollections, straining almost to aggression and downcast almost to self-effacement at once, his eyes intent on pulling free of their orbits and rolling down his face or, failing such acrobatics, begging to be noticed in the athleticism of their effort so that an onlooker, feeling a sympathetic ache or blurring of vision, might serve up to him delectable descriptions of the auto-erotic spectacle upon which he had become fixated without once getting a fix, dish up his own lips as if they were a meal in a formal social arrangement for whose purposes a slap would serve as well as a kiss, both kinds of contact swinging appetite toward enjoyment along the wet kinetic trail of what he called, indifferent as to its precise meaning though anything but indifferent to the aura it arrogated to his own rather less industrial-strength jones, the smack. There, in something a bit larger than a nutshell, you have the secret early life of a man named Birdy, for the retrieval of which, if I have had to enter into a series of relations perhaps too close for comfort and too strained for critique, I offer as amends the consideration that *I have adapted it even for the purposes of our social relations, transposed it into the form of politeness which we favor and which endeavors to be positive and precise, to approximate itself to the plain truth.*

CORRESPONDENCE

It becomes a woman's picture. A memory palace taken apart by a rabble of editors. Xanadu. *The interrupted. The unfinished work. One way to remain unfinished is to stop. The other is to go on.*

—*Laura Moriarty,* Nude Memoir

...a way of notating rubato.
—*John Tilbury, "On Playing Morton Feldman"*

...an utterance at climate level
—*Mark McMorris*

Tell her – the page I never wrote!
Tell her, I only said – the Syntax –
And left the Verb and the Pronoun – out!
—*Emily Dickinson*

1. [Eds.] The letter attached here was not addressed to the editors, and thus makes an uncomfortable fit with which to open a section of "editorial correspondence." Nonetheless, address and arrival are two different things – the same can be said for several other letters in this section.

Thursday the 30th
Under the Kumquat Tree,
Behind the Galvanized Fence

Dear Two Gray Sisters More or Less,

I have *not* renounced my political affiliations, and to tell the truth, the name you've colonized isn't even dangerous or clever enough to piss me off. Furthermore, I am not scratching around your door and whispering through your vents at all hours, much as you would like to interpret a sentence like, "I don't know, I have to shop for cat litter," given in answer to your question, "When will you shatter the horizon within which your taking away of the expansiveness of my breath has imprisoned you; when will you scatter the cops who line the barricades of humid air and cinderblock between us; when will you return the world to me?" as if it meant, "There are vast landscapes of hot and poisonous fertility against which my body must yet rub and slink before it can assume its true proportion." Yes, there are certain synthetic musics that reverberate backward as well as forward without apparent regard for time or narrative ontology, and for a while I did sleep with a crazy woman who played in such a band, ten years ago, maybe more, I'm not sure but I'm not the one laying claim to the very mechanism and meaning of memory here, and in any event this does not equate to my body becoming at once unstuck and static like the ferro-concrete recording angel of our era, guarding an open tomb in a sealed archive in the expansive interstices of a cramped acoustic ecology, against which illusion, covering and eventually displacing my plain speech with the *plein air* romance of ambience and the dim buzz of participation, I suggest you switch off the stereo and actually take a listen or two to what transpires outside your window.

That it's an old, old story isn't exactly the problem, for *I see no objection to its being old, but whatever else it is it's not* euphonious. The couplet you scratched in the wet cement of the new interstate overpass reads like marching orders, as if at the global scale of such address my many selves, or the many selves to which you have recourse in your amateurish attempts at an arraignment, might cohere into an avant-garde *only* a lap or two behind the pace car. In fact, you seem at best to hold a kind of merely historical interest, like the giant birds that used to bring us water in their ornate beaks, and clean towels stashed behind a door or flap in their downy bellies, and that became irrelevant rather than extinct, serving as markers of a program, too half-assed by more than half, for the structural transformation of the public sphere into a hypermodern rotating torus decorated incongruously with lots of dark draperies and wax fruit platters. You still see them sometimes – the birds, not the platters, which got eaten when the welfare state collapsed again – frozen to the benches outside the old DMV office, waiting in vain for it to reopen. In that sense, certainly, I am more than willing to congratulate myself on my ability to navigate a sidewalk or two without waiting for an airport to be built according to a union-busting "enterprise plan," or for an up-to-date transit map to be deployed, and then the after-hours addenda, and then the several errata slips, and then, and then.

As far as I can tell, the whole fairytale of your pedagogically intertwining arrested adolescences to this point amounts to a long-winded, backhanded, strenuously easy confidence in how much you'd like some sort of monster to fuck, how monstrous you find the very thought of fucking anything that's not a monster, and how intent you are on droning on and on about it by any veiled means at your disposal. Like all those old stories, *it ends just in time, but it ends badly!* Sometimes you experiment with a new arrangement of facial hair and I think the difference in pronunciation a mustache makes might take the whole thing into unanticipated angles, but no, it's the same tired, tried and true crime-scene triangulation, over and over again, shot with rubber bullets so that, barring the occasional mishap, everyone gets up and does it again on the next shift. Except, of course, for the assistants, a category that you force to grow larger with every passing turn of phrase, and into which faces pass, never to return into recognition, just as others rise up out of that mass, parting the long grass down by the drainage culvert and walking onto the scene as if an artifact of some hideously complicated butterfly-effect of a wind that blows murdered or murderous innocents across your path like dandelion seeds. In that, there's a hint of the sacrificial genital carpet with which you were so taken, "seminal" and "foundational" blending in grassy overlap, packed down to bear up the image of a used sedan you'd have me drive so that, saying its name, you recall the specifics of my route at any moment, but for which, given that high-mileage domestics break down at regular intervals along the way, you have to allow these wide and repeated margins for manufacture, test drive, the years during which the car is "pre-owned," and a full staff of extras – those assistants, again – for each phase in the general production. What complicates matters even more is that you want something like unforced approval, a ratifying plebiscite enfranchising every one of the participants, as if they might *hold out their hands with a hesitant air that was meant as a proof at once of the inculcated reserve which they had to overcome and of the spontaneous friendliness which successfully overcame it. Your voice* – yes, I know I'm the one writing here, but really, you're not in a position to insist on the distinction, after all the sleight-of-hand that's gone before this moment – *has risen to a series of high notes on which it rests as though awaiting something, holding on to them in a prolonged expectancy, in the exaltation of already seeing the object of its expectation approaching, and with a desperate effort to last out until its arrival, to welcome it before itself expiring, to keep the way open for a moment longer, with all its remaining strength, so that the stranger might pass, as one holds a door open that would otherwise automatically close,* the faint attention paid to a downy and ornate fringe of décor, an interior of gratifyingly scratchy fresh hand towels, failing to apprise you of just how out of date your struggle for the heart of the moment, that center ring full of dusty, once-fantastic circus animals, really is.

Anyway, *my hand trembles* under the allergic strain of all the floral excess with which you've decked this scene, making me sick by way of miming concern for an illness that was never mine. So I'll be brief and bright, hot and toxic, like the sudden flash of ignition from the reflective bit of gas line you were accustomed, until this moment, to regard as a rehearsal partner as you cobbled together a language of public gesture from the out-of-phase overlap of degraded, private caresses. Twenty cops beating children in the vacant lot down the block don't make a sound you can imitate in the affirmative, you know.

"It is in the air, at large, like a pure and supernatural being that unfolds its invisible message as it goes by." You *will* persist in giving names to the weather, and to inarticulate grunts incidental to politics by other means, and thinking thus to have named and accounted for, as Apostrophe, as Twenty Times Yes, as Americana, as Mom, a series of women – various, uniform, as if buttons themselves could press an artisanal hand into the service of color-composite screens and dodges – for whom you remain, to be blunt, a traffic incident yourself, a flooded local street necessitating, at most, a quick stop to remove one's socks and rub the distributor cap dry. *Doubtless the form in which you have codified those*

charms of local color, the scum of decaying oak leaves, pine needles and straws from takeout soft drinks floating, damping down the ripples on the surface of the tannin-stained summer runoff, as the storm drain chatters happily in the near distance, almost drowning out the sound of the hail of blows, as if to drown were somehow a more fitting form with which to shore up nostalgia, or as if the whole musical ensemble, hail of blows superimposed by additive synthesis upon hailing frequency, might cast nylon filament after nylon filament into those waters and fish out at first a tangled mess, later to become a mesh, a fabric for a form-fitting, all-occasion sort of leisurewear – doubtless, I was saying, that form *could not be resolved into rational discourse,* even as it founded what would come to count, for you, as the very possibility of such a language. *You know that the piano falsifies still further the perspective in which you see the elements of music* – too much like typing, in which reduced ballet, for you, typography always gives way to typology, never passing through its resemblance to the piecework your mother did incessantly, to the point of injury, during one depression or another.

I began this letter expecting to clarify a point or two, bring you up to date on the mundane specifics, speculation aside, of what I've been up to over the decades you two (or 1.*n*) have played at calling to your account, but now I see that to do so would only lead you to the bridge above the cresting river, transcripts and tape recorder in your hand, thinking to submerge the age and its "recording angel" (your bit, words in my mouth from my pen – a nice touch, though wrong) in the fluvial wash of amnesiac, ultra-specific enumeration. Doubtless, you will interpret this as the final, charming flaw that renders me real flesh and blood, skin against whose hue you might check, as if to correct, your own, and whose reservoir of salt your tongue would strain to relate to the tears by which, you'd come more and more to believe, you had called me into being. The upshot of your cathartic sneeze and snuffle is a swarm of mosquitoes thick as description. Already, citizens are fleeing the one metropolis the Gulf Coast could muster and barricading themselves behind overturned pressboard tables inside Airstream trailers deep in the north county green belt, the regimented ranks of the slash pine plantations phasing in and out depending on perspective – now a bit of forest, now a bit of architecture, on and on. Before you say it, no, I didn't catch the bug that started poles and wires sprouting from everyone, and yes, it's incurable. I know you read in the car, smoking a cigarette with the window rolled halfway down and the wet August afternoon sliding its airs around the planes and curves of your face, all sorts of occasions for distraction and disaster, which from the other lane, head-on, looks like distraction *from* disaster, and though you've invented a car that stays in one place and forces the world back and forth on the force of whim, I must tell you the net result looks the same, not only from where you sit, but anywhere else as well. What I mean to say is, pay attention to what I'm telling you, but keep an eye on the road.

A few points of interest, curiosities that occur to me along the way: you had entire cities cryogenically frozen so as to keep track more easily of the populations into which I might come and go, and what did you call me? You rented automatic dinosaurs, given an evolutionary reprieve in the service of shoddy simulations of a new extinction, to uproot and chew apart the boltholes in which I might have kept a low profile, calling them eyesores and drug dens by way of calling yourself rightful executor of the rights resident in the plots of ground beneath, and what did you call me? You suspected yourself of keeping a secret file marked with your own name, and implanted yourself as a mole in that operation, but in that sea of cross-indexed, color-coded alphabet tabs, what did you call me? You voted my slate without question, convinced that this would lead to a relaxed conversation in loosening clothes over slowly warming cans of beer, but what did you call me? You set out bear traps in the driveway, hoping to catch me as I siphoned gas from the used sedan you'd repossessed, and when you heard that first bone-shattering snap, just before you realized it was a member of your own landscape crew you'd caught and maimed, what did you call me?

When you stop *imagining what's kept secret from you in the light of what's revealed,* call me. Until then, let's call this a business arrangement.

From the middle distance,

Epistrophe

2. [Eds.] For an explanation of the receipt of this letter, see #1 above. This would seem to be a reply to that text, likewise received, mistakenly or not, by us.

Friday the 31st
Against the Wall,
Above the Background Hum

Dear Epistrophy,

So *you're not made of marble,* after all, and the Monk reference leads me to believe you've "come to your senses" after a spell. Following that "off-minor" string of *meaningless and dangerous jokes* at who knows whose expense, riffing along the sign chain as if all the world were an exercise in paronomastics, "Let's Call This" absolutely anything at all, apparently – well, what a relief! Having read the salutation and glanced at the signature, I'm sorry to say I let the letter slip while opening the car window to flick a spent butt into the median strip, but I can imagine well enough the rest of the head-shaking that went on in the body of that wordy missive (too much to read during the commute, even had it stayed in place against the steering wheel, so I must be forgiven for cutting to the chase), your eyes refusing to light on any single outcropping in the landscape of a past from which you hasten away, full speed ahead – and who would blame you? – while the blur of that back-and-forth of denial brings into impressionist focus before your eyes, taking on specifics flatter, subtler, more specific, the contours of a new peninsular landscape within which we might begin anew together, in which I take you by the hand and say, "If only we'd known we could sew, connect the dots of impression into this press of flesh and atmospheric pressure with that characteristic sea-level weight."

Perhaps now you are ready to thank me for my forbearance, the reserve and patience with which I have *sat here like a surgeon waiting for a spasm to subside that has interrupted my operation but will not make me abandon it?* In any event your letter, fluttering out the window into a humid late afternoon, against whose stillness my forthright and determined velocity had created a whipping wind, made a sound that said, "Yes, yes, twenty times yes!" I seem to remember having heard that somewhere before, and take it as a sign that you will give up your obsession with rooting out these nameless, faceless and, might I add, entirely imaginary assistants with whom to strike up the worst sort of opportunistic political liaison at every turn, even as I take it up as a sort of aria, a command performance all my own in a road movie soundtrack, with insertions:

Call it out of line.
Back that unit down
and out of range. Pure domain
opens up pure domination.

"Am I trying to hurt you, captain?"
The stick held in the belly, point
blank urge self-managing. That,
plus hours of all the saddest songs.

Fire. Hungry women.
Lovely children. More
than any other town

more elbows turning in
the air around her tuning
in tangible static all
about him inspires you
the audience with wild
traffic what's that sound
that's or what's escape as
it whirls away or as it
surges back into your view

past the convicts one can't
allow to get involved with one

more sordid prison's reason or sense
or force of some conviction. One's
own conviction. One's double sense
of balance. These are no "results."
Youth detention facility shoots by
behind fifteen-foot cyclone fence.
Free movement of late-model cars. All angles
of incidence without incident. Blank or void
"resultant" of those bug-sized forces.

"How did we do yesterday?" he asked.
"We made the miracle of touch typing,"
and his lower lip trembled, tongued surplus,

squirting you
back at you.
I'm going
to die anyway.
I don't care

if someone's following me.
But the magazine man *does* seem
to bring around a new subscription
every time we move. Well, then I started running

snapshots from *A Photographic History of Song*
in a special four-part series. He was sickeningly
disappointed. Sick and hot with migraine, feeling
inverse sympathy for all those headless creatures.
He never wrote a fragment. Extracts actually
grow inside the reader's brain, like cockroach eggs.

He is in your ear. Your car.

His language. "Cast iron."

Describe medieval music abstracted from the festival,
the festival today abstracted from the downtown square:
free blues in the bank plaza. Didn't we pass this town
before? Sound installed with ringing blows inside

the quintessential early metal building

money dancing to refresh itself

calculation running hard before design

— *pleasurably, without shyness*
— *loud, the second time round*
— *the game goes back to snares*
— *once more, with feeling*

You wake in cramped quarters. A billboard flashes past:
"You have a son-in-law in the Department of Accidents."
Like a friend in Jesus, or Pennsylvania. Bumper sticker
the negation of a license plate. Polemics in the passing lane
with short-haul freight: sewer sludge, cable spools, tons
of brick, under tongues of men and women, still themselves
among the bubbles, in a pool of pork-rind-flavored
spit, mortaring a discourse into auto-presence.
Pull off up ahead. Who wants another bag of these?

Anyhow, you *will* persist in turning something as simple and endearing as a pet name into an accumulating coriolis of wind, sighing from my slightest syllable only to intensify, miles away, into the storm blowing from paradise, or if not paradise, at least the reinvigorated high-speed port just behind your writing desk, full of the promise of bulk-ordered kitty litter, flashing phosphors, thick volatile beauty of data clouds, and the volunteer prosthetic civic labor for which in some belated present tense you, newly specific to a particular and organic body as the utopian payoff of that binary stream, anticipate how best to register your thanks, as if for the inordinate transparency of space itself – this wind, wherever from, that catches your fingers as they clutch at ethereal orders of presence and renders them suddenly slack with playback, your repetition and my reconfirmation of the subtly insinuating smiles that exempt you from the series of family monologues and place you in the sender-receiver chain as an acceptable other, the same wind, in fact, that powers the surgically-enhanced birds by which our benevolent grasp of natural default in the palm of husbandry has provided us with hygienic dispensation from the upper atmosphere itself – and *because* you will persist in this detour of the simple nickname into a multiplying complex of dependent clauses, I will find it, from this moment onward, loveable, and will *wallow voluptuously in the emotions* it carries between us as if they were scents borne across the sea between two continental landmasses, fires and dust storms in Africa giving the South Florida air a tang of almost identifiable longing that precipitates a thin layer of dry powder coating every surface along Interstate 275 with a bit of surplus magic in which I might roll and against which I might rub, were I not so intent on reaching the destination promised by your sweet and inspirational missive, proposing, as it seemed to be about to do before an uncooperative atmosphere intervened, a new and nearly automatic expansiveness and expansionism for our devotion to each other, assisted at every turn and junction by The Buddy Nation, Inc. Robot Labor Pool of Automatic Romance, vital hydraulic fluids humming thickly under their polymer skins like sweet cream in which a low-speed beater turns.

In other words, if you want to make out with girls, or birds, or factory machines, that's alright, since now I know you're only seeking me by other means, and after all, I can't be everywhere at once. What I'm trying to say is that there's a whole atmosphere between us that has your name, and that *there's never been any moonlight to touch it. My eyes remain fixed upon the things I do not know.* For that reason, if for no other, I take my inability to get a clear prospect on how my own body fits into this as a sure sign that the whole bureaucratic brain-sharing apparatus has seen its last days, that my other fraction has slipped back to the future, a last shared song on both our lips, lilting, *"When somebody loves you, it's no good unless they love you all the way."*

The pain was so intense I was obliged to stop.

Sign me,

All of Me

3. [Eds.] The following reproduces the text of a letter received following the first installment of the serial publication of the manuscript in several newspapers. We have remained faithful to the letter in all particulars except as regards the physical state of its text, which was originally set in what is popularly known as the "ransom note" style. Considerations of space have led us to include it here as a print facsimile.

Dear Editors,

I have been trying to get through by phone. You will not be addressed, apparently. Just as well. I have no proper face with which to address you. I am both stupid and innocent, you know.

Nonetheless, I must insist that an accurate representation of these circumstances *is possible,* even in default of any mutual or one-sided recognition. Perhaps because of that default. The foray into criticism on the flipside should go some way toward the simultaneous establishment of the archive and cancellation of the plate.

Sign me,

The Skin Donor,
Or, Riffling Through a Ream of Nastiness

[Eds.] The following text was found on the reverse of the letter, and running on to the front of a second sheet.

Silky-smooth, they'll come
to call me now, or

child once my despair
found out its danger—
additions accumulate

instead of piercing straight
through short lapses in
this thinking far too long.

Never fear after
so light a moment
should blame you much
as stacked odds piece

by piece, she says, became the same.
Show what sainted names you please:

—ostensive definition—

—voluntary fruit of force—

a pistol followed home
the red-eyed course pretends
its purity lulled the I askew,

the single loosened "and"

to collate alibis page by page.

The innocent world flutters eyes awake
to liens found sole among a crowd,
foreclosure on its leaning plot
of habitation, habit, dirt.
That author's page lacks
all discretion, and history,—
that greatest, and that very worst

4. [Eds.]The following was received in typescript, postmarked from a location some several thousand miles distant from that of the previous letter, on the same day. We assume, therefore, that the writer is referring to a previous letter, known to him or her, which never managed to get through to us, and not to letter #3, of whose text the author of #4 must have been ignorant. The architectural references in both the prose and verse sections of this letter support an assumption that the text is, like #3, intended as a critical response to a passage in the manuscript – in this case the phrase "whose characteristic architectural operation was the removal of anything at all characteristic." Here, however, we lack the certainty of the previous letter's direct citation from the manuscript, and our assignment of this paratext to any particular location in the text should be taken as speculation, however well supported by the available evidence.

Dear Editors,

Your previous correspondent was on to something, notwithstanding the put-on. Something was trying to get through. Nonetheless there are rooms, entire buildings even, given over to the interminable. Perhaps it is from some such place that I address you, in the middle of a labor capable only of segmentation, never an end, never a

present tense:

Desemanticized,
these buildings eat their context after all—

however the development of theories about a city called *Benevolent Knowledge,* which resemble flies from a distant dimension, an unorderable, heteroclitic dimension, might like their *things* arranged in such American *outwards.*

Many thronèd Powers,
essences embattled,
perishable in stoppered
bottles: for the happy State

that swallowed up our conqueror.

(Now the phone's hollow ring
yields up my unspotted soul, fuzzed
with line noise of corruption).

But shall I subdue translation, or,
forward to prevent such horrid fray
as more than Heav'n permits, know
troublesome taste, and be henceforth

every knowledge or opinion
of retiring, sky-tinctured walks?

Or go forth, accompany us our own
complete receding from the day

we shall die
and all reasons that
not eating satiate
at length

— to sapience, distinct
forbidder,
all spies at last
will come —

or jump, you dancer, bodied slack enforcement
rubbing closure on the lids, running up and down
some sacred obelisk or other like a slip
of native tongue

5. [Eds.] The text of this letter is written on the reverse of #2 above, making it clear that this is in fact a reply, and raising certain questions regarding the identity of its author. We can certify that we were in receipt of letter #2 continuously during the day on which letter #5 purports to be written – and is in fact postmarked. The production of alibis, however, is too far outside our job descriptions, and thus we offer no further explanation.

Monday the 3rd,
Under the Street,
Just Above the Threshold

Dear Three-Quarters Full,

I thought you might be saying something, i.e.,

"I'm Hit. Go On Without Me.

With these names one shows that one's burnt paper one handed with one's one hand to one's one-handed hammer in one's pocket with one's other one hand for one's protection on one's late-night walks from one note or one's notes in the margins of one's music – *not* "one music." Which makes one two: the colorful dead-end sentiment *and* action in the same act by which definition is lacking, or is lack – *not* "one lack," though not not that, being not this – e.g., that there is no description, no described, no describing and no circumscribing the difficulty with notions such as 'imaginative re-enactment'. Is this the *Symposium*, the trick trauma phenomenology of a men's movement urging me to mourn an unremembered circumcision, English as a second mortgage, or the name of your fairytale tacked on to your letter to the editor, all of which I've ripped off here?

Something deemed impossible because matter: "Winter followed past life within limited areas of self-contained economy. Problems of recurring civilizations largely understood in terms unproductive and related to involvements more mere than maudlin and intoxicated, falling into curbside slush." Though that as well. All these signs according to which "radical cuisine" very early meant militants were precisely the absence of the most difficult piece of brightly colored glass. The basic impulse more than any other seems forcibly coordinated. The experience that form regained for freedom soon lapsed into shudders

of social and scientific consciousness, intellectual energy of the permit and permission slip. A whole lot of stuff slips in, actually: body politic, body politics, body popular as it offers comfort to an uncomfortable and overstaffed bureau of the human condition, be it not much more. For an artistic representation is, no matter rich depiction, a knowledge. There can be no embodied hope exceeding what was first poured into pylons holding up the pier

as strong as her physique: broad, flat little ears. A lean, blue image appearing at the end. Hologram or ghost? Reflection made reception true, that he had failed that message precisely when he understood, *but the passive cells overrode my faction.* One glance at the heads-up display calls up the command: Drive fast with the top down, cut the forward section loose. The prow plows a pure and stupid wedge on through the surf, while cargo founders on the sandbar far behind. High and dry, we two disclose to sand and grit the rounded folds the rain rubs raw."

...but when I drove by your place, your door was open and your mouth was closed in that lipless lock you have such a way with. There is a name for the pain that holds you in that posture, if you really want to know, and for the geography whose features it connects and so subsumes. But it seems your every communiqué these days arrives by misdirection, or by hydraulic robot pump, or pimp, or must be salvaged from the fetid mess disgorged across the landscape by one of those high advisors with whom you will, of course, disavow any relation more complicit than having tuned in once or twice to the flickering screen of his address, indulging in a bit of self-laceration by way of payback for dropouts in the playback. When they return to you, calling attention in red ink to your failure to clarify the terms of your own address, do you *cast them aside, then cradle them again in your bosom, like corpses in a river?* Has this all been academic for you, an exercise in false memory starting from the fact that the wrong person's recollections are at stake? There are people living in the storm drains as corpses, who pick through your remains and construct a chain of echoes to be sent back into thousands of tract houses, and if you choose to call this news of the world, ignoring the obvious mathematics in which something slightly less than two makes something slightly less than whole when split according to the dictates of the most equitable distributive utopia, so be it.

Right back at you,
Antistrophe

6. [Eds.] An interlinear response to #5 above, this text also turned up in our possession.

Tuesday the 4th,
Beside Myself,
On Top of the World

Dear Empty-Stuffy,

I am an endless string of corpses holding hands, passing in review before the painted-shut window in your head, each fixed rigidly in its posture and incrementally distinct from the one before so that, since the road runs on a fairly steep downgrade past your apartment, forcing one, if one were one and not a daisy-chain of dead men, to wonder whether one, or the many nones of one, is or are in Florida anymore, and since there's always money to resurface the roads, despite the temperate climate and the relative permanence of anything here that keeps a low profile, one or the multiple teleological negations that point to one gathers or gather quite a bit of speed passing by, flickering past in fact so that what you see is a lively dance across the landscape where really there are only a bunch of stiffs. It's all staged for your benefit, you see, this *brutal and improbable fondness* of the human flipbook that amuses you so. These other guys aren't really my friends, simply the means for me to make clear that, identical but for small intervals of frozen gesture though we might be, they don't know you like I know you, and therefore one, as an exception to one's nones, might say they know nothing, are worse in fact than a bunch of those robots, the vital fluid basis of whose stiff forward-straining movements constitutes a veiled threat and implication of our obsolescence, if not quite *mine*.

I anticipate a day when high society will hold fast to the grip bar on the bus and share the social drama, marked for special distinction only by the feathers of the obsolete Hygienic Delivery Birds in which it will be decked. There will be a particular magic mirror before which all shall pass in review, renewing their belief that what the frame contains is in fact the arrangement of reflection's fragments into a colloidal blob, which the heat of all our collective excitement can be counted on to bake into hard, translucent shape, and with belief thus renewed, each in turn shall mortify the voluptuous suspicion that what the filigreed wire encircles is nothing but a zone of empty air before the leering face of the assistant who holds up the ill-tempered, parodic object as if it were an accusation. Walking by that bit of glass or trick air today – for my least utopian impulse, no sooner thought, is carried blinking and still wet into the street – I noticed that my reflection looks like you, if you were more kindly disposed and more expensive, not, I must add, through the arbitrary value-added mechanisms of fashion, but through the application of a real cottage industry's craft aesthetic.

But how in the world can I see what you're doing, when I'm thirty-five years away? The feeling that I now entertain for you, being no longer tinged with pain, could scarcely be described, now, as love, though I must confess that, as a dead guy shambling around in the radioactive glare of the badly colorized early evening, the thought of my teeth parting your hair and sinking into the warm buttery folds of your brain (I think that was the image with which your poem ended) makes me almost giddy, nearly a blip or two above the flat line with desire. In any event, if people ask, be sure they know that *I* dumped *you*, as the chemistry of embalming transmuted me to something different in kind, infinitely more precious, long before I reached the quarry or sinkhole in which *you* dump the bodies you truck around like so much poisoned weight of gypsum tailings.

Stiffly,

Mr. Bird-in-Your-Hand

7. [Eds.] As we were preparing to send off a second installment for newspaper serialization, we came across the following passage, in what looked to be a fair-copy version on a single sheet inserted at this point in the manuscript. We are all certain – and photo reproductions made only hours before confirm us in this – that the inserted page did not accompany the text as we received it, and must have been inserted while we were out to lunch. Nonetheless, the hand is an exact match for that of the manuscript proper. Could this be the work of the anonymous author? Absent further confirmation, we have opted to include the passage here among the "Correspondence," it being in some way a communication addressed to us in our role as editors, though to what end we cannot say with any conviction.

PLEASE INSERT THE FOLLOWING:

Playing the recording backwards at double speed later that night, I made out the voice of a little girl, breathlessly explaining, "That you persist in thinking, despite your proximity to the nerve center of the whole thing, on speaking of these wars as events, rather than evacuated spaces, voids into which future events can be dropped, makes me want to weep and hide my face from you, and so I have."

Thus, when a later president told us that he felt our pain, I knew exactly what he meant, since from that childhood moment onward it was clear to me with a crystalline precision that such pain would have been of necessity rigorously tested and sampled in advance of its application to our bodies, or to the bodies that might have given us our own bodies by refusing to be the bodies we had, were they not entirely consumed as fuel. Grandma Violet muttered something ambient and sad about deserted beaches to the south, where we might live as pure respiration, and leave our respirators behind. Mr. Article created counterpoint and crosstalk, intercepting signals from a public radio sponsor whose company organized human nature hikes so well-disposed they hardly required any walking at all beyond the initial wrench of dislocation, from *me* to *I*. *He seemed to have penetrated my father's skull, and to be seeing at that moment, far beyond and behind it, a brightly colored cloud which provided him with a mental alibi,* though the trees continued to speak for someone kept studiously offstage in these after-dinner skits, intransigent and wooden.

[Eds.] As if this were not enough difficulty for a single day's work, a shift in gravity that wrenched us out of the position in which we had hoped to recognize ourselves when we took the collective job description of "Editors," we received a rather unsettling phone call later that afternoon. Having only moments before faxed this section of the manuscript over to the newspaper which had accepted it for publication in its Lifestyle section, and being that moment on the phone with its editor confirming receipt of the pages – minus the previous insertion, whose provenance remained too uncertain to admit of its inclusion without extensive editorial apparatus – someone rang through on our private line. In a voice obviously disguised by means of a heavy cloth or other membrane held over the mouthpiece, an unidentified party dictated the following text to us, specifying that it would be of some use "in your present dilemma, that is, the dilemma of the present tense. Here is a face without attributes to accompany a text for which you can't produce an attribution." The newspaper editor was chattering happily away on the other line, ruling out the only other person who knew at that time about the uncertain insertion. Even he had not seen the text, having only that moment heard from us a general account of its arrival. Was this then the voice of the manuscript's anonymous author? The question remains a present one, in answer to which we can only reproduce, as closely as the garbled voice allows, the less than helpful text below.

Battery gets leashed to fall a route forward,
but stealing shape, a coda etches
out ascent. Too long
down to reserve reversible liquids,

the second's up: broken showdowns of state.
The rising rate of patch-ups.
The racy window. The gaudy hearts talking
in their proper slots. Let me sing you,

two yous, a shred. As wet
could so run out no accent who
wink billfolds in raw trance, lashing firm
to radical pistons, illusion for the dense "who,"

for little else asserts another rate
of zeroed pellets of doubt.

8. Dear Editors,
I fell in love like that, once, at the mall, a grain of sand wet upon my tongue and a viscid floater in my eye. It gave to the object of its speech a title,

the well-positioned honorific of nobody's fool.
The well-remembered little fink you were.
The saturated thickness of the food-court air.
Idling. Legitimating high in neutral. Then quantities
of time elapsed – say, movie-length. Slow focus
of the adjectival qualified some names among us,
killed a fixed percentage of those engines, random
outburst of small business folding into history.

"Even in the public-sector cold, I lost none of my sense
of him, his risk, that torque and hum and burnout
heat arcade

and piercing kiosk were hard-pressed to contain,
condense, retail."

"No one else I've met came to me wrapped
in bouquet-blasts of quite such credible detail!"

O violets! O eight-bit graphics of the violence
blooming all around beloved heads!

The air of fate had vanished after final cards were dealt out on the screen. And tracing the complex chemical reaction from that point on, while doubtless possible, would simply produce an order of description on the same scale as the events themselves, like the footpaths marked along the floor for early morning power walkers. One might as well try fighting the oily, inflammable tide, blanketing the turbulent seascape with a rich retardant foam of ethical discourse and didactic good intentions. But it was my blood was tainted, after all, and all my circulatory metaphors for the coming heart-to-heart compromised from the moment the big glass doors spread open in such unqualified welcome.

Despite these qualms I deeded over all my tiny flames to him.

The haste of the dissolve recrystallizes in the genitive,
material, genital. The friction of each fractional rub flakes off
another solid body from the spirit capital of your selfsame other.
Here's genetic criticism of the image of oneself: over there

inside those tight and clinging jeans.

Yours,
The Editor

9. [Eds.] The appearance of this bit of portraiture, printed rather than written and to all appearances free of water damage, baffled us for quite some time. Questions regarding its authenticity plagued us almost until the very moment the manuscript was set to go to press. Just as we were on the verge of dropping

the final, corrected proofs into the "out" box with this inconsistency still unexplained, a package dropped into the "in" box. Without salutation or return address, it held only a too-dark photocopy of an anthology of Gulf Coast Florida suburban folklore. The second article in this anthology – whose title page, unfortunately, was not included in the photocopy variant we received – is a new translation and commentary on the traditional tale "Thumbnail Real," beginning with the handwritten version of the poem in question. We reproduce it here, intact in all particulars save the poem, which differs only in handwriting from the text presented on page 40, and the commentary, which is directed to a specialist audience far removed from that intended by the present publication.

Once upon a time the sex workers of a certain city formed a political party in alliance with the hotel and restaurant workers' union, and issued a decree requiring the rich to eat and defecate, one at a time, in solitary exposure in the midst of public squares, traffic circles and malls, while a representative cross-section of the city strolled by, refusing to take notice. A family called Sodplug, who refused to take anything at all seriously, put on their best only-on-Halloween, heterosexuals-do-drag-too outfits, and went out to meet this scene, whereupon their faces cracked apart, revealing vast distances between what they meant and what they meant to mean, and forcing the invention of velcro.

"Such interesting people, and we love to watch you get it on," they chanted into the bedrooms and hotels, occupied at that moment with an odd and unaccountable array of bemused people in evening dress, caught on film in the act of choosing among the many different styles of forks.

"When you're a storm from Paradise, no one gives a rat's ass about your opinion," they chanted in unison, though most emphatically *not* in chorus, lifting up the metal plates that covered the trenches in the street. This is where they found a certain Mr. Bird-in-Your-Hand, primping before a mirror, or at least a reflective bit of water conduit or gas line, a magical bit of heavy metal that showed him to himself as he must appear, swaddled softly in the burning hydrocarbons that set his features off to such effect, during his odd and unaccountable public appearances.

The mental image of the heaps of heaving flesh, spanked and pinched into a hot glow hovering just between health and trauma, which would greet him every morning as he stuck his head from the flap of his shamanistic pup tent, no matter at what distance he had set his government in exile from the centers of population, never failed to arouse in him an aggrieved principle of aesthetic moralism, even as it simply aroused him.

[Tutti]: Thumbnail real...

[Birdy]:
...is the sorrow that gives words' sulphurous end[14] –
counseled career her blur
angling acute or grave by others' means
she'd keep forgotten in the distribution
of acoustic space in cells
seeds, flowers, functionalist speech
neither well paved nor well built.
Murder measures stairs' descent
by nude grasp or signal's noisy generality
bicker glance aside.
Sold body reflected to itself
the dole of kitsch collectibles
Post-surrender electronics ready
detail for the mobile occupation

— made up to open wide,
too much spreads too thin too soon you
big blood sugar bully, you.
The same examination minus
too much space's tortured calm —
the voyeur wired to its chaperone,
the necessary genital remark —

[14] [Eds.] Was he dead of hard radiation, or the *"I don't know, I don't open invitations"* of his love, all scale plan and names?

remarkable mass transit rhythms and
around-the-clock laughter of police
accommodate this promiscuous image
neither on the nose nor right between the eyes.

Homogenizing serial force allows
the arming of mechanical arrangement:
I saw you then I saw you
fall then things helped me remember,
and then a moment's blank, hot to fill,
does magic to anxiety
of tiny bodies' now,

the ordinary craze or commonsense rapport in wake
of 4am earthquake or noon opinion poll, the voting
hand steadfastly in the fire.

Which is to say that Bird-in-Your-Hand listened to the panting, wet breath of a bed partner as much for what it might tell about the current state of post-national flows of power and capital through the unwitting bodies that had come along before his to touch the body upon which his earthy witticisms now played, as for the simple gratification and confirmation of his own complexes. I have it on authority that my own parents – for this all happened long before I walked onstage, and thus I must adopt the mode of the fabular in the telling – during this time refused Birdy – for so he was called before taking on the mantle of a thorny and stoned oppositional dignity in more recent years – the simple pleasure of acknowledging that they knew anyone other than himself, Lenin, and Richard Nixon to exist, and found a sadistic glee in thus forcing his natural impulse, which was to spread himself out, thin and clinging, over the entire turbulent surface of his society, like an oil spill in the Gulf of Mexico during a hurricane, into a narrow and deep channel like a semitropical mockup of a fjord, through which the shipping was altogether regular, the docks collectivized in exemplary fashion, and only the Enemies List admitted as the single representative instance of their own incoherent lateral spread, the homeopathic means by which this implication in processes and circuits not entirely salutary might be deflected into an inclusion whose conspiratorial totality was so overdone as to leave them in fact high and dry on the banks of a new pastoral of the left out.

"There is nothing vision can reveal that does not run away into a suspect idealism," Pop was fond of saying in those days, according to what Mom has told me since, "and so it's best to squeeze those eyelids down tight, before you risk an eyeball popping loose as it strains after what's best described as a metaphysics of absence."

Birdy responded to these conceptual difficulties by pretending to be oddly and unaccountably sexy to the young married couple, a near-distance into which their bodies might move so as to gain a view of themselves in all their made-up nakedness, when in fact he was up to something quite different, subtler, which involved duping their attention into unsuspecting focus on the brightly-clad and musky figure he cut in the sensorium of she or he whom their political allegory of vision rigorously avoided noticing, thus making of himself a wrench in the works something like, to use an anachronism more proper to my own time, the guest who brings a tray of sausages to the vegetarian potluck. But then, as one of his lovers had been fond of saying, had anyone been of a mind to listen to her during the barbecues at which she was invariably a guest, and just as invariably kept behind an epistemological blind as one keeps the hole knocked in the plaster during the last party hidden behind a potted plant when the landlord visits, it was certainly true at all events that Birdy made a point of living in neighborhoods in which there were no unironic potlucks, because there was no exchange, because nothing at all spoke, and all the houses turned to face the street and the promise of mail. This, coupled with the permeability lent his conceptions by the several holes in his skull – he being, in fact, the earliest victim of a cannibal freak named Mr. Article, shortly after that person's emergence from the wreckage of a certain ideal of density – and despite these very holes and the method of their imposition on his adaptive flesh and bone being evidence that something in the landscape was very much in the habit of getting close and whispering to him – this muteness and this porosity were what allowed him, nonetheless, to develop his unique conception of a legible and tangible history, available for critical inspection to him who would only once consent to rub his holey head against the ground and let the dirt add grit and realism to his brain.

10. [Eds.] The reader might be interested in the following text, which transcribes a message left on the editors' answering machine following the serial publication of several sections of the manuscript and during the time in which we were busy preparing this section for a similar form of publication.

Hello? Hello? Well, since I can't get through I'll have to leave this on the machine. You know who it is.

Crackle as handset is placed near a radio and a station tuned in. Broadcast begins, an excerpt from an unidentified opera.

[Tutti]: That independent teenage radio projectile…

[Solo soprano]:
That greatest and worst of
    interactive irony machines
    each body to reflection as its home
Phase-shifted afternoon on the lawn
            curbside   alter   oilslick
                        as if wound round its mouth
                        and your ass will follow
Calisthenic countdown   Shades' anthemic wraparound

[Solo tenor]:

Spackle compounds this
fleshy rhythmic smack   large mass
accelerated upward by the sensitized medium
notate temper   music made
uncornered   Package Drive-Thru
Expect trajectory   closed curves
transact a delicate head   terrified
affective arbitration of the phone-pole posters
...highly unstable second
calmly and quietly slackens
— behaviorist attractor...
legible...   topiary mess —

[Solo soprano]:

Furious...   a million years inferred
among the flowers...
survey method settle
to the building's floor   beaten
to resemble distant ancestors
dug our roots   *unworked dirt*
discords   *that*   to every   *there*
— melts into the central air —
The Man hatches friends —
change of address
speaks —   *your disappearance*
but "in person"
in agony   longing for the same
scene obscured in smoke

[Duet plus boys' choir]:

The law explaining all
self-abducted   abdicate   apogee

recalls an instrumental curve *all the*
little children up on stands
lead the murder ballad chorus

Fadeout. Radio announcer gives the time and station call letters, but does not announce the title of the performance.

[Eds.] After several calls to the station and a personal visit, we have determined that no opera was included in any playlist for that day. The scheduled program for the time slot in question was a discussion forum on proposed zoning changes.

11. Dear Editor,

Once upon a time, you could drive through the city stoned and create a makeshift eclogue out of that fuzz of vague participation, much as earlier drivers had created their pastoral investments with property in rolling parkland. In the windows which were not plywood painted to resemble glass, but behind which were walls painted to resemble all-encompassing skies, people would perform the parlor drama of their social obligation to your passing. Going through someone's trash, or the stack of papers on someone's desk, or the pages inserted by a printer's error into someone's typescript, was a hot and dirty intimacy. The substance of a whole affair might take root in a similar sense of euphemism.

Ours,

The Editors

12. [Eds.] The rare appearance of a date on this page enables us for the second time to connect the typescript to another text, this time an e-mail received earlier the same day and saved on the anonymous author's hard drive. The bulk of the message's memory size is taken up by an attached graphic file – see the reproduction below of this scanned, handwritten text. The subject line reads: "RE: In absorbing my dreams, you have delivered me from them," and thus assumes the appearance of a response to an earlier message. Unfortunately, no such message could be found on the author's computer, or on any of the various disks stashed in the desk. Likewise, a single sentence of body text reads as if a conversation picked up already in progress: "Here is that doubly-anonymous portrait of me as the just and accurate representation of you." The

return address points to a cancelled, free webmail account. Readers are encouraged to make of this what they will.

contradictory at seams the swelling
armature from arms embraced return fire
to a cold lump of ballast the battle
so under brick the tongues that lick
that sing while you're at it that aggression
might imply just time a stake
in the heart of the matter
of facts' ass-backward bloody drag
a series of interminable terminal
states bordered in congested tissues
imply a just time's stake in
your singing at aggression while it might

yet turn that weight to throw
bolt into public works away at you

13. [Eds.] The following is the transcription of part of a tape-recorded interview with Mr. _______, the "real world" prototype for Mr. Bird-in-Your-Hand, conducted by a local television news show running a "decade in review" nostalgia series. It was waiting in our VCR when we returned home from the editorial office upon finishing work on the final section of the prose narrative.

...I was so drunk I could hardly steer. Something crystal clear ended up embedded in someone's scalp. Now, I thought, at last he or she – whoever – would understand me and what it was I dictated into the old reel-to-reel tape recorder I'd picked up at the pawn shop down the block, homebound by the used sedan which, having changed hands one too many times, could no longer be aimed in predictable parallel with the center line by force of hand alone. In other words, *my meticulous prudence was defeated one evening when I had gone out to a party.* There followed a period during which "Uncle Madame" (we still can't use the real name) and I worked at building our own model society through primitive multitracking experiments, in which a tape made our duet a trio and then a series of multiples, filling the cinderblock box with a growing chorus of voices saying the same words out of phase and allowing us to posc as linguists, hearing a new language in the way staggered old syllables gradually mutated into struggling new phonemes, as the attentive ear slowly shifted from the center to the edges of articulation. *It is a society as a whole, now that I am detached from it, which presents itself to me as a series of pictures.*

14. [Eds.] We received this letter while preparing the galleys of the text for publication as a book – that is, following the final installment of its serial appearance. To judge from its salutation, it is misaddressed. To judge from its tone, it would appear to be a further response to letter #6 above. The lapse of slightly more than a year between letters, however, calls that reading into considerable question. We offer it here as we received it, without further comment.

Wednesday the 5th,
In the Air,
Up to Here

Dear Zombie Flight Crew,

I know it's loud back here in coach with the hippies, dictators, and ghosts retreating from the future mass deaths of Buddy Nation, Inc. Prosthetic Robot Mnemonists, and yes, I know my voice is slurred around the quantities of overpriced and watery domestic beer I downed while waiting, in the international terminal's cocktail lounge, in the wake of the crushed uprising, to begin my career as an alias, and yes, I know it's a long way forward to where you sit, navigating into all that backwards-blowing wind, but really, I've hit the call button several times now and the least you could do is make an announcement. There is someone wandering around back here and I think he belongs up there with you, steering or something, not accosting the sore-throated overnight passengers with his *eyes welling with affection, ready to detach themselves like tears and fall upon his face,* and once detached, ready to ascribe themselves as easily to any other passing face, whether it wants to own another set of bleary and reflexive eyes or not. Perhaps you need to fit him with a more distinctive style of comic-relief operatic helmet, so he'll recognize himself in the scrape of horns on the low cabin ceiling when he goes abroad in the cabin looking for adventure. He has just now started a fire in the lavatory compartment, and is staggering blindly through the smoke among the rows of seats, pinching arms and screaming, "Thank god, it was all a dream!"

"So manifold are our interests in life that it is not uncommon, on the self-same occasion, for the foundations of a happiness which does not yet exist to be laid down simultaneously with the aggravation of a grief from which we are still suffering," I am telling him, having read it somewhere, and knowing, now that I've had to write this up like an incident report to pull a bit of your collective attention away from your headlong flight into the inevitability of what's just so, that you'll soon have read it too – have to have read it, in fact, this penultimate chapter in my treatise, *Matter and Memory and Memory:*

Gravity. His head pulled away from his skull.
Pooled resources of medical shutdown
eased the field flat. *We need to find*
equivalents to ovaries.
My research grant made me say that.
Some structure that had pressed against me *here.*
The gradient is increasing, I think. I think:
the spiked obscenity. A sense of corruption
ran like fluid indeterminations of
her shape as *emptiness* proposed in his new
arrangement of the molecules one did not want:

the water-carrying hump, flow channels
through which a hot flush and cold filter could
scavenge sorted elements, ferrying need
along the downhill wave, back up to the pump.

By light or radar, bits of units lifted off the ground
onto the screen, held by force that comprehends and cramps.
Sleight-of-hand reels out the high-speed motion study
slightly prior to events. I know precisely how much
of your saliva ends up trickling down my throat.
Intimate experience redesigned in fractional form.
"I should be a single ripple washing down the line."

Stretched skin, beaten on a chillingly familiar frequency. From a single point half a degree past crazy, surging figures and expanding vocabularies want to agree to anything, to look past almost-historical incidents into the perpetual motion machine for which I hereby give you the schematics, count out for you the pulse.

"You're right about that – maybe you *could* help me slip into a more complex dialogue than the smoothly overlapping scales of a face, doubled at no matter what angle, in no matter how many intersecting bits of one-way glass." The tangled structures in his brain hummed as another set of heads drummed staggered rhythmic cells upon his thighs. Somewhere and somehow and shadow, and he was invisible to himself, and we watched him from the observation room.

His eyes ached

closed by force
of will against
the plainclothes heat
of cheap terror.

The bulb. Her
shoulders. A blur
of sweat. Her
syllable, binary: "one."
Meaning: "on." Light
reveals the mirror,
makes it "no."
No light, no
heat. No sweat.

****

Whatever,
End of Something

P.S. The bird has flown beyond the point at which I could or would care to make some sort of appeal to your moral instincts, but let me interrupt this broadcast and make it clear for what it's worth that I am perfectly capable of groping around inside myself, and have found, in so doing, that you are not the mouth into which you have been putting words by miming being spoken through and spoken for, but that our relation springs from each interval in the annulment of that string of falsehoods, and every time you close your eyes it comes to light again that I am nowhere near dead enough for you.

Shake on it,

Auntie Sophie
a.k.a. Auntie Terrible

ERRATA SLIP (MAP AND MODEL AND MOTEL)

> *How could he breathe?*
> She inhabited her brain,
> filled it, squeezing him
> like gum against the wall.

Seeing the bind in which his prospecting had placed him, she held his hand around the automating filaments, and gave a solid yank. If the nervous impulse translated into motion, it was one too fine for her to see, or him to feel above the pounding blood in his ears. She froze, staring down at her legs as if willing them to walk. And his body flopped over, all alone at the center of a vast circle of bright light that had cleared the dancefloor when his drunk stagger had first made itself apparent.

Was he dead of hard radiation or the *"I don't know, I don't open invitations"* of his love, all scale plan and names? These latter are plateaus crowned with stretches from the larger building projects, the scattered ruins representing single houses. The ancient landlords in their vast proportions, stretching far toward extinct

in their hundreds

even thousands of

screens and screenings for barbarism in the sequel when it first discovered North America by way of the North American drive-in. We were perforated to pass the land and camp, passing land within and through our perforated flesh so that the camps we pitched would always have surpassed the land while claiming to be simply "passing through," perforating it like light through apertures playing on a thousand screens.

> We call these bluffs. We pass them during bus rides.
> Or buttes called outriders. Called out as pastimes,
> each segment calling out the line. Call it out of line.